£ 3.00

I̱A̅

Industrial Archaeology

A SERIES EDITED BY
L.   T.   C.   ROLT

7

The Chemical Industry

The
Chemical Industry

W. A. Campbell

Longman

LONGMAN GROUP LIMITED
LONDON

*Associated companies, branches and representatives
throughout the world*

© *W. A. Campbell 1971*
First published 1971

ISBN 0 582 12799 8

*Printed in Great Britain
by W & J Mackay & Co Ltd, Chatham*

Contents

List of Illustrations

Line Drawings in the Text

Sources of Illustrations

Nos 1 and 2, Muspratt's *Chemistry as applied to Arts and Manufacturers*, London, 1860; No 3, Parke's *Chemical Essays*, London, 1827; Nos 4, 5, 6, Lunge's *Sulphuric Acid and Alkali*, London, 1886; No 7, Richardson and Watt, *Chemical Technology*, 1863; Nos 8, 10, 18, *Alkali Inspector's Reports*, 1864; Nos 9, 12, 13, 23, 24, *Penny Magazine*, 1843, 1844; Nos 14, 17, *Newcastle Evening Chronicle and Journal*, Picture Library; Nos 20, 21, *Technical Records of Explosive Supply*, 1915–19; Nos 11, 19, Company advertisements; No 15, *Daily Mirror*, 1906; Nos 16, 22, 25, Author's collection.

Introduction

The history of the chemical industry is a story, on the one hand, of chemistry developing against a background of geography, engineering, economics and sociology, and, on the other, of personal triumphs and disappointments of a number of great men. The tale might therefore be told in a number of ways, according to the viewpoint and qualifications of the author. In this book the emphasis is on the nature of the scientific problems which manufacturers have had to solve in a variety of circumstances. The heart of the matter is that a chemical reaction which is feasible in the laboratory may be hopelessly difficult, or totally uneconomical, on the factory scale. Upon that transition from laboratory to plant fortunes have been lost and won and reputations made and broken.

Best thanks are due to Professor N. N. Greenwood of the Department of Inorganic Chemistry in the University of Newcastle, in whose laboratories and with whose encouragement this work was completed: and to Joan Stafford for painstaking assistance in the preparation of the typescript.

CHAPTER ONE

Prelude to Industry:
Salt and Copperas

When Richard Watson became Professor of Chemistry at Cambridge in 1764, he confessed that he 'knew nothing at all of chymistry, had never read a syllable on the subject nor seen a single experiment on it'. Nevertheless this remarkable man (who later became Regius Professor of Divinity and Bishop of Llandaff) was able in only fourteen months to prepare himself to deliver courses of lectures to very full audiences consisting of persons of every degree. Before relinquishing the Chair of Chemistry sixteen years later he wrote his *Chemical Essays*, in the five volumes of which he described manufacturing processes for green vitriol (copperas), alum, nitre, common salt and sal ammoniac, and gave details of the extraction of lead in Derbyshire and the smelting of zinc in Bristol.

Here, then, is evidence that several chemicals were prepared commercially in the British Isles about the middle of the eighteenth century, and on such a scale that a man who had approached the study of chemistry with an open mind was quickly made aware of the processes. In fact we can extend Bishop Watson's list considerably, for we know from the writings of Peter Shaw (*Three Essays on Applied Philosophy or Universal Chemistry*, 1731) and Robert Dossie (*The Elaboratory Laid Open, or the Secrets of Modern Chemistry Revealed*, 1758) that Epsom salt, Glauber's salt, oil of vitriol, charcoal, acetic acid, and the pigments of lead and iron were all made in quantity. To prepare the way for the great chemical manufactures of the nineteenth century we shall examine in this chapter two of the chemical enterprises that flourished at the beginning of the Industrial Revolution.

Salt

Any material that binds itself into the language, customs and folklore of men must be important to them, and salt has done this more than any other chemical. The tallest story goes down better with a pinch of salt, waverers have the horrid example of Lot's wife to keep them forward-looking, the baronial table was divided between guest and serf by that same salt which causes so much dismay when it is spilt. It is not surprising, therefore, that saltmaking in this country should have its roots in a very distant past.

The primary commercial outlet for salt was the preservation of fish and meat throughout the winter, a sheer necessity before the revolution in agriculture that enabled livestock to be kept alive until the spring grasses returned. Before the Dissolution, the great monastic houses as well as the castles of the nobility had either to purchase large amounts of salt or make it for their own use. The grim Abbess of Whitby (whose tale is told in *Marmion*) had salt pans and so did the powerful Abbots of Durham and Wearmouth; indeed communities of saltmakers owed their continued existence to these ecclesiastical patrons.

So far as Britain is concerned, there have been four methods of winning salt: by boiling down sea water, by evaporating salt springs and wells, by digging it out of the ground and by pumping it out in the form of artificial brine. The first method is also the earliest. Sea water contains about 3 per cent of salt or roughly sixty pounds of salt to the ton, so that given a supply of cheap fuel saltmaking became an attractive proposition. The presence of coal near to a sea coast provided the right setting, but also brought the salt trade into disrepute. Probably the earliest complaints about atmospheric pollution arose in connection with the salt pans. A petition to Parliament in 1617 claimed that 'such abundance of thicke smoke doth rise from the same pans as all or most part of the grass growing upon the ox pastures within twenty score yards of the top of the bank next to the pan is altogether burnt up and wasted'. A century later, when Lord Harley made his *Journeys through England* he wrote 'the houses [of the salt workers] are poor little low hovels and are in a perpetual thick nasty smoke'; the workers themselves he judged 'brutish and devoid of decency and religion'.

Chemists will wonder why smoke should destroy grass over such a wide area, and the reason is revealed in the reports of the Alkali Inspectors from 1863 onwards. Salt from the boiling pans often dripped on to the hot coals beneath, when the silica in the coal ash caused the liberation of hydrochloric acid gas.[1] For as long as brine continued to be evaporated in open pans, this problem remained unsolved. The coal itself was not without blame, for only the poorest coal—not worth the cost of transport and otherwise likely to be burned at the pit-head—was employed at the salt pans; this grade of coal was known as 'pancoal'.

The mode of manufacture was as follows. Sea water was collected in a sump, situated between high and low tide levels and in which some preliminary settling of organic matter could take place: from the sump it was pumped into the pans, which were topped up from time to time as the volume decreased. The salt water was not evaporated to dryness but only to the point at which crystals formed; the crystals were fished out, leaving behind a solution of calcium and magnesium salts known as the 'bittern'. The pans were made first of lead and then of iron, and were of sizes from fifteen by ten feet to thirty by twenty. They were worked in pairs or in fours with a central gangway between them for stoking; sometimes one pair was elevated somewhat above the other so that waste heat from one pan might be used on the next.[2]

The initial settling did not, of course, remove all plant and animal matter, and in order to bring this to the surface so that it did not contaminate the salt each salt master had his favourite additive. William Brownrigg's list of 1749 mentions wheat flour, resin, butter, tallow, new ale, stale beer, wine lees and alum, but he says darkly 'some boilers prefer the fat of dogs'. The most usual clarifiers were egg-white and bullock's blood, though even here personal idiosyncrasies enter, for one writer held that the bullocks should be black. Whichever method was adopted there was likely to be a stench and a disposal problem. Perhaps this is the reason that North Shields (a salt town in Northumberland) was described as 'an unhealthy town, refuse accumulates in the streets, the only scavengers are the pigs and plague is a regular visitor'. The coal ash also presented a waste disposal problem and in Durham County houses were built on ground levelled with ash.

The principal sites for this kind of salt extraction were the Firth of Forth, the Ayrshire coast (Saltcoats), the estuary of the Tyne (the major English site) and Lymington in Hampshire from which the considerable naval requirements for salt pork and salt beef could be satisfied. The salt was transported from the pans by way of salters' roads, and the geography of the trade can often still be traced in street

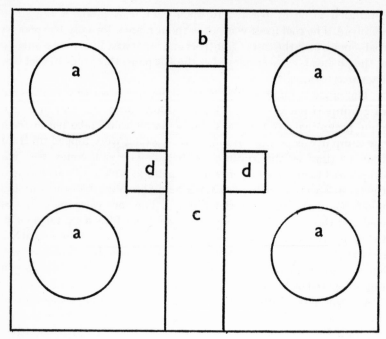

Figure 1 Layout of a Saltern
a. Salt pans
b. Fuel store
c. Working area
d. Fire holes

and place names. Many districts have names ending in -pans, Prestonpans, Howden Pans, and these are often linked with commercial routes or population centres by such names as Salters Road, Salters Gate or Salters Bridge. The charting of these salters' paths is a fitting occupation for industrial archaeologists.

Before leaving the subject of sea salt it is worth noting that at Lymington, on the edge of the New Forest, salt was actually obtained by solar evaporation, a process which we are more likely to associate with the Mediterranean than with the Solent. According to Celia Fiennes (*Through England on a Side Saddle*, 1698) evaporation by means of the sun's warmth was carried on for sixteen weeks in the year. From September work began on boiling down the brine

obtained from sea water during the summer months; for this purpose Newcastle coal was brought by sea, and it seems likely that a
trade based on this more expensive coal could not have survived without the fuel saving due to the solar evaporation. The Lymington
salterns reached the peak of their prosperity about 1750, when fifty
thousand tons were being produced in a year: a further boost was
given to the Lymington trade by the production of Epsom salt from
the 'bittern' to the extent of four or five tons for every hundred tons
of salt.

The evaporation of sea water was altogether too slow a process to
provide a raw material for the chemical industry. In any case, the
trade was in severe decline by the end of the eighteenth century,
partly due to the discovery of methods for coking small coal so that it
could no longer be had for the asking, partly due to fiscal pressures,
and to the availability of more cheaply produced salt from Cheshire.
At the mouth of the River Tyne, which Sir Willian Brereton (1634)
had described as making more salt than any other place that he had
seen, only a handful of pans remained to serve some specialized
requirements of the glass trade; of these, one was kept in use throughout the nineteenth century for recrystallizing rock salt and was
finally cleared away in the 1920s. On the east coast of Scotland, too,
an industry which had exported largely to Holland in addition to
filling the sacks of the Edinburgh salt-wives was virtually extinct by
1826. The great Prestonpans works finally closed down in 1959. The
last of the Lymington salterns was converted to an oysterbed in 1865.[3]

A much quicker process was the evaporation of natural brine
springs, for these are usually about eight times as strong as sea water.
The most famous of these was Droitwich spring in Worcestershire,
referred to by John Leland in his *Journey* in 1538. This salt is
associated with gypsum or calcium sulphate which, being only slightly
soluble, tended to incrust the pans and cause waste of fuel. It was the
custom at Droitwich for descaling to be carried out about once a week
by a man who waded through the warm brine with his feet in two
wooden buckets. The pans became bigger until 1880 when they
reached the size of thirty by twenty-two feet and one foot nine inches
in depth: such a pan when new was deeper at the sides than in the
centre to allow for flattening out in use. There were also sizeable
brine springs at Stoke Prior in Worcestershire (at one time worked by
William Gossage) and at the 'wiches' in Cheshire. It was, however,

the discovery of rock salt in Cheshire in 1690 which led to the modern British salt trade.[4]

About the mining of rock salt in this country there is little to say. A shaft similar to that for a coal mine is sunk, and the rock salt is broken by blasting. Because there is no gas in a salt mine, the elaborate precautions against fire and explosion that are so necessary in coal mining are unnecessary here. The rock is so firm that galleries can safely be made broad enough for comfortable working. The most celebrated salt mine in England is the Marston Mine to the north of Northwich: salt was blasted by gunpowder, sent up in tubs and shipped down the River Weaver to an extent that reached 110,000 tons in 1882. Nevertheless the mining of rock salt has never satisfied more than a small proportion of the country's salt requirements, for rock salt usually has a reddish discoloration and needs further purification before use. Most of the salt consumed by the chemical industry has therefore been obtained by dissolving out the rock salt in a current of water, and evaporating the resulting artificial brine. The two principal sites for this trade are also two of the main centres of the heavy chemical industry, Cheshire and Teesside.

The Cheshire brines were evaporated at temperatures varying from 60° to 100°C, to yield salts of different grain sizes: the finest salt came from pans which were brought to boiling. The pans were arranged in series of three, and fired so that the flues from the two outer fires met under the centre pan; in this way two pans could boil while the third was maintained at a lower temperature. The pans were made of quarter-inch boiler plates riveted together and set on to a framework of brick. As soon as salt began to form it was scooped out in perforated wooden scoops and put into wooden moulds of square cross section: these were allowed to drain for twelve hours, and the salt was then tipped out and dried in a room warmed by waste heat going to the chimney. Once a week the pans were roughly cleared of gypsum incrustation, and once a month were cooled and thoroughly descaled with hammer and chisel.

In 1863 rock salt was discovered at Middlesborough on Tees by Messrs Bolckow and Vaughan the steelmakers. The deposit was covered by a bed of hard rock which made the prospect of mining somewhat unattractive and the project was abandoned for a time: meanwhile other deposits in the same area had been discovered by Bell Bros, who were also ironfounders. In 1886 three of the major

1 Sal Ammoniac Plant, Kurtz, Cropper & Co., Liverpool, 1860
2 Manufacture of acetic acid, Beaufroy's Vinegar Works, South Lambeth, 1860

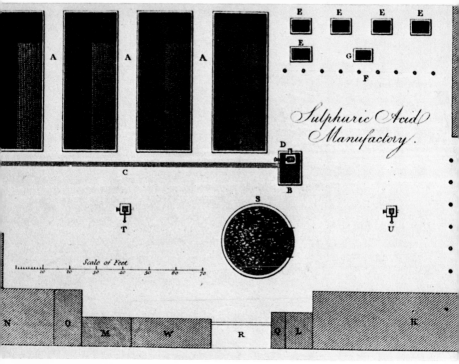

3 Plan of sulphuric acid works, 1827

4 Arrangement of sulphuric acid works, 1879

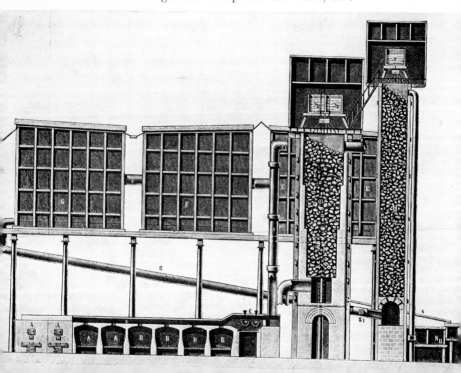

Tyneside chemical manufacturers joined in a salt winning exercise and brought over American borers who were experienced in drilling for oil.[5] Boreholes of five to eight inches in diameter were drilled and lined with steel tubing (the earlier Cheshire borings having been lined with wood and clay). Salt was brought up as brine and evaporated in pans of enormous size, some being up to seventy feet by

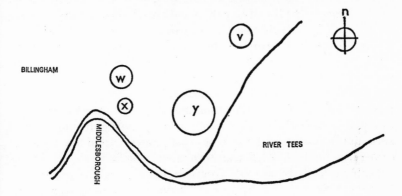

Figure 2 Major Salt Borings in South Durham, 1888

v.   Allhusen
w. South Durham Salt Co., Haverton Hill
x.   Tennant
y.   Bell Brothers

twenty-four. As changes in the method of alkali manufacture took place, solid salt was no longer required and Teesside brine was used as it came from the well (p. 52). The protecting layer of rock has saved the Tees salt areas from surface collapse as salt is dissolved out: Cheshire, where the salt was much nearer the surface, has been less fortunate and subsidence around Northwich, Middlewich, Nantwich and Winsford has been the cause of much litigation.

Just before the end of last century George Weddell, a Newcastle associate of Sir Joseph Swan of electric light fame, began experiments on the influence of small quantities of phosphates on the running properties of table salt. Having achieved success on the laboratory scale he set up a works at Haverton Hill on the northern edge of the Teesside salt deposits and there commenced the manufacture of what is still known as Cerebos Salt.[6]

Copperas

Due to accidents of etymology, the science of chemistry has acquired some perverse terms; theobromine does not contain bromine, fluorene has no connection with fluorine, and copperas is not a compound of copper. The misleading name stems from a centuries old confusion between two kinds of crystals which—from their likeness to pieces of coloured glass—were known as vitriols. One of these, blue vitriol, contains copper, whilst the other, green vitriol, does not; it is easy to see how this coppery title has slipped from one to the other. Copperas, green vitriol, ferrous sulphate heptahydrate, $FeSO_4.7H_2O$, occupied a key position in commercial chemistry from the mid-eighteenth to the mid-nineteenth century, although its occasional use in Britain goes back at least two centuries earlier.

Copperas was made by the atmospheric oxidation and hydrolysis of iron pyrites, sometimes called 'fool's gold' from its glittering appearance but more often known as 'coal brasses' from its frequent occurrence in coal. This transformation was formerly known as 'ripening'. As early as 1693, Matthew Falconer of the province of Brabant in the Low Countries was known to be making copperas in this way at Queensborough in the Isle of Sheppey. In 1678 Daniel Colwell contributed to the *Philosophical Transactions of the Royal Society* an account of copperas making at Deptford: he described the pyrites as 'copperas-stones' and said that the brightest specimens were used in wheel-lock pistols.

The Deptford copperas beds were about a hundred feet long, fifteen feet broad and twelve feet deep, made watertight with beaten clay and rubble: a wooden trough laid in the floor of the bed conveyed copperas liquors to a cistern under the boiling shed. Small pieces of pyrites were laid to a depth of two feet and left for perhaps five years for the ripening process to complete. To afford some degree of continuity to the process, fresh stones were laid on the earlier layer after about four years: in this way a bed did not become exhausted until it was filled to ground level. The cistern at Deptford was made of oak boards, well caulked, and could hold about seven hundred tons of liquor: there were transverse bulkheads in the cistern so that a leak developing in one part did not entail the loss of the whole batch. Rainfall was essential to the reaction, but artificial watering systems were not recommended.

Pyrites, FeS_2, contains more sulphur than the formation of copperas requires, the result being that copperas liquors contained sulphuric acid. The density of the liquor was tested by 'swimming' a boiled egg in it, and it was noticed that good liquor would dissolve the shell of the egg in three minutes. Partly to take up the excess acid, and partly to preserve the green colour, pieces of scrap iron were placed in the cistern: without this the green ferrous salt would have turned to a yellow ferric sulphate. The liquor was boiled down to a small bulk and run into a cooler in which twigs and branches were suspended. As crystals formed on the twigs they were lifted out and dried.

The main purpose for which copperas was intended was the production of ink or of black dyes, particularly for hats: there was indeed an Act of Parliament in 1565 which forbade the dyeing of black caps in any other way. Most vegetable extracts allied to tannins will produce dark colours with iron salts, but the most favoured ingredient of inks and blacks was an infusion of Aleppo galls. In dyeing, the fabric was first boiled in a solution of galls and then steeped in copperas solution. The knowledge that copperas could darken the colour of vegetable infusions was put to nefarious use in the 1850s. Used tea leaves were collected from tea rooms and coffee houses, dried, mixed with a little gum and copperas and sold again as fresh tea. The small amount of tannin still remaining in the old leaves was sufficient to give a dark coloured infusion on boiling. Establishments for carrying on this trade existed in London and Liverpool, and when one of these was raided by excisemen in 1847 they found sacks of copperas stored in the back rooms.[7]

The large expansion in the manufacture of copperas came about through the appreciation of new outlets for easily obtained products. From 1750 onwards copperas itself was less important than the materials that could be made from it, and since the processes of decomposition required coal the trade developed in the coal areas: the siting was, of course, helped by the fact that pyrites was often an unwelcome byproduct of the coal field. Thus Faujas de Saint-Fond, the traveller and geologist, described with admiration the scene of Tyneside in 1748.

The industry of the inhabitants of Newcastle is so active, that accustomed to apply it to everything, they have even turned to profit the pyrites which injures the quality of coal but which is found in abundance in some of the mines. The pyritious substances are carefully separated from the coal, and the expense which this labour might occasion is repaid with usury by the vitriol which is produced'.[8]

The first Tyneside works was opened in 1748 at Hartley by the Delaval family, who had coal and glass interests. This was followed in 1772 by a large works at Sunderland, and in 1789 by the Walker works of Barnes and Forster which continued in operation for a century. Thereafter copperas works sprang up on both sides of the Tyne along a length of about ten miles, much of the product being exported to France for use by the dyers of Rouen, Lyons and Marseilles. A similar development was taking place in the Scottish Lowlands. In 1755 a Liverpool company began to make copperas at Hurlet where it later became a cognate of the great alum trade there. Other Scottish works at this time were started at Campsie and at Stirling.[9]

When copperas is strongly heated an oily liquid is produced: this is the 'oil of vitriol' or concentrated sulphuric acid. Left behind in the retort is a residue of red iron oxide, variously known as jeweller's rouge or Venetian red. At best, one hundred parts of copperas can yield thirty-five parts of oil of vitriol and twenty-nine parts of Venetian red, and it was upon these two products that the later prosperity of the copperas trade depended. The sulphuric acid was used to prepare Epsom salt from magnesian limestone, and alum from alum shales: both of these found their way into the dyeing markets as mordants. Epsom salt for medicinal purposes was also prepared from the bitterns of the salt pans. Venetian red was a pigment for the paint trade, a fact which seems to have influenced the way in which individual copperas businesses developed.

Because the transformation of pyrities into copperas took several years, those engaged in this manufacture were forced to operate by-trades, more or less connected with copperas. It seems that the Venetian red outlets took most of them further into colour making, such additional enterprises as oil-milling, turpentine distilling and the making of lamp-black being common: moreover, as will be seen in Chapter Eleven, many also made Prussian blue. This throws difficulties in the path of the industrial historian and archaeologist, for the same firm may be listed in successive directories under a variety

TABLE 1   **THE COPPERAS TRADE**

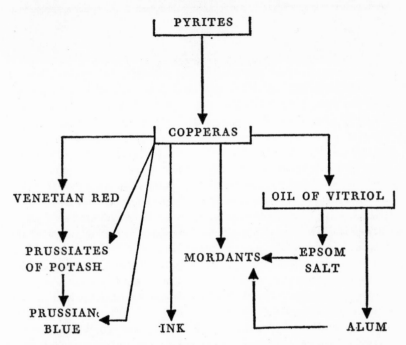

of business classifications. Table 1 shows the ramifications of the copperas trade.

Two factors contributed to the decline of copperas making. The first was the discovery of better ways to make sulphuric acid, which we discuss in Chapter Two. The other was the development of synthetic dyestuffs and pigments to such an extent that Prussian blue and the other iron colours were toppled from their pre-eminent position. The copperas works that survived into the middle of the nineteenth century were on a very large scale indeed. When the Felling Copperas Works in County Durham was valued for sale in 1855 the buildings, yards and beds were found to be worth £4600 and the lead in the cisterns and vats was estimated at £2691 as scrap.

CHAPTER TWO

Sulphuric Acid

The pronouncement of Liebig, that the commercial prosperity of a country might be measured by its consumption of sulphuric acid, is quoted too often to need comment. Familiarity, however, should not dull our appreciation of its truth, for over a period of some two hundred years sulphuric acid has been a foundation upon which many manufactures have depended. In the eighteenth century, tin-platers, japanners, refiners, bleachers and makers of mordants relied on it: during the nineteenth century it sustained the two giants of the chemical trade, alkali and dyestuffs, and today it is used in making fertilizer, rayon, cellophane wrappings, detergents, titanium dioxide pigments, in treating sewage and in filling batteries.

We do not know who it was who first made dilute sulphuric acid for pharmaceutical purposes by burning a mixture of sulphur and saltpetre under a bell-jar in contact with water, and this is not the place to attempt an unravelling of the twisted threads in that story. So far as preparation on a manufacturing scale in this country is concerned the priorities can be established with certainty. The date was 1736, the place Twickenham, and the man Joshua Ward. Ward's career in the field of proprietary medicines is dealt with in Chapter Eleven, but it is perhaps germane to the present issue to remember that, following a shameful episode in a parliamentary election, Ward had spent fifteen years in France, where he might have imbibed some of the chemistry of Lemery and Geoffroy.[1]

Ward's Twickenham venture seems to have depended on the technical skill of his assistant John White, described in a pamphlet of 1763 as 'the Ingenious Chymist who carried on the Great Vitriol Works at Twickenham'. The plant consisted of two rows of round-bottomed flasks each of about fifty gallons capacity, partly buried in sand and so placed that the necks were horizontal. A charge of sulphur mixed with nitre, contained in an iron dish and standing on a small earthenware pot, was introduced into the neck of each flask and ignited by touching it with a red-hot iron. Each flask contained a small quantity of water, and the neck was closed with a wooden plug. When the combustion was complete a fresh charge of sulphur and nitre was introduced and the process continued until the acid was judged to be of sufficient strength to justify concentration by distillation. It is said that the process caused a nuisance and that Ward was forced to move to Richmond in 1740, but there is no sound evidence for this.

Gabriel Jars, a French metallurgist who visted English factories, alleged that Ward guarded the secrets of his process by employing only Welsh-speaking workers. Whether this is true or not he delayed taking out a patent until 1749, by which time others had moved into the field and overtaken him. Although the chemistry of Ward's process had been described by others before him, three small but significant adjustments to the mode of operation seem to justify his position as the founder of the sulphuric acid trade in England. The form of the apparatus allowed the charges of sulphur and nitre to be renewed without disturbing the dilute acid inside; the arrangement of the flasks made it possible for a man to be gainfully occupied for the whole of his time, for when the last charge had been ignited the first flask was ready for recharging; and lastly the scale on which the operation was carried out represented a definite transition from laboratory to industrial conditions. The success of the venture can be gauged from the fact that sulphuric acid fell in price to something between 1s 6d and 2s 6d per pound, or about one-sixteenth of its former cost. Readers of Boswell will remember that when Dr Johnson bought an ounce of oil of vitriol from a chemist's shop at Temple Bar in 1772, he paid one penny (a halfpenny less than he expected).

Lead Chambers

When the patent of Ward and White was filed in 1749 John Roebuck of Birmingham had already been manufacturing sulphuric acid for three years. Roebuck, born in Sheffield in 1718 and educated at Leyden and Edinburgh, had originally intended to practise medicine in Birmingham, but joined with Samuel Garbett in a refinery for precious metals in Steelhouse Lane.[2] In 1746 they began to make their own sulphuric acid, substituting for Ward's glass globes what Roe-

buck called 'lead houses': he is thought to have been influenced by
Glauber's statement that sulphuric acid does not attack lead. Three
years later Roebuck and Garbett established a second works at Prestonpans, eight miles south-east of Edinburgh, where there were
already considerable salt-making interests. Here the lead houses
measured 8½ feet by 6 feet by 4 feet and thirty of them were accommodated in one room: they were firmed against the wall, strengthened
with wooden framing, and bedded on sand for steadiness. They cost
£8 each.

Ward's iron pots and earthenware dishes were replaced by lead
vessels which were introduced into the chamber through a ten-inch
diameter hole fitted with a lead stopper. Each charge consisted of one
pound of an 8:1 mixture of sulphur and nitre ground together, and in
the bottom of each chamber was five hundredweights of water. After
burning had continued for two hours, the hole was opened for an
hour before a fresh charge was put in on top of the old one: when
three charges had burnt the 'salts' were scraped out of the lead dish.
A month was required to bring the acid up to the strength at which
concentration might usefully be carried out (about sp.g. 1.25). In
spite of what seems like tedious slowness, the Prestonpans works
established an export trade to France, Holland and Germany in
addition to supplying users in England and Ireland.

Having relied on secrecy for over twenty years, Roebuck decided
to patent his lead houses in 1771, but by this time a series of leakages,
betrayals of trust and piratical thievings had spread the details of his
methods over a wide area, and two years later Roebuck went bank-
rupt. The first factory in the London area ever to use lead chambers
was founded at Battersea in 1772 by a firm of druggists, Kingscote
and Walker. Their chambers were cylindrical, six feet in diameter and
six feet high, and they had seventy-one of these in addition to four
cubical chambers. One of the Walker family established the first
Lancashire factory near Manchester in 1783 and by 1792 there was a
flourishing works at Bradford. Table 2 indicates the spread of vitriol
making by 1820.[3]

So far we have considered the making of sulphuric acid as a raw
material for sale to users. The really significant advances in the
trade, however, took place when the acid began to be made by its
users, bleachers and soda makers. Bleachers used sulphuric acid to
liberate chlorine from a mixture of common salt and manganese

TABLE 2

LIST OF VITRIOL MAKERS IN ENGLAND IN 1820

Brundrum & Co.,	London
Farmer,	London
Smith,	London
Hill,	London
D. Taylor & Sons,	London
Liddiard,	London
Dobbs,	London
Skey & Bewdley	Staffordshire
Caves & Co.,	Bristol
Bush,	Bristol
Dobbs,	Birmingham
Austen,	Birmingham
Phipson,	Birmingham
Paton,	Birmingham
Bower & Sons,	Leeds
Norris & Son,	Halifax
Betson,	Rotherham
Doubleday, Easterby & Co.,	Newcastle
Rawson & Sons,	Bolton
Do.,	Bolton
Watkins,	Manchester
Mutrie & Co.,	Manchester
Do.,	Whitehaven

(James Mactear, *Proc. Glasgow Phil. Soc.*
13,1881, 409.)

dioxide; the chlorine was then used in the form of an alkaline solution called Eau de Javelle (Chapter Six). In 1798 Charles Tennant ('Wabster Charlie') took out a patent for a solid compound of chlorine and lime, known as chloride of lime or bleaching powder, and for the manufacture of this material he set up the St Rollox Chemical Works in 1799. At first he bought sulphuric acid from Prestonpans and other factories at a cost of £60 per ton delivered, but as consumption in the bleaching powder works grew he erected lead chambers in 1803.

Tennant's chambers, six in number, measured 12 feet by 10 feet by 10 feet and cost about £50 each compared with Roebuck's at £8. The chamber house was a three-storey building occupying an area of 50 feet by 24 feet. The chambers were mounted on the upper floor, the glass retorts for concentrating the acid in the middle, and the lead pans for the initial concentration at the bottom. Between 1808 and 1811 there was introduced a most valuable improvement in the way of working the chambers. Previously the sulphur-nitre mixture had been burnt inside the chamber, and a good deal of both chemicals went to waste in the 'salts' or 'sulphur ashes'. Starting from an attempt to burn the sulphur ashes more completely, an external furnace was fitted to one chamber in 1808 and the system was extended over the next three years until all the chambers were so served: one furnace supplied two chambers, the direction of flow of the gases being controlled by dampers. A further step in the improvement of the process was made in 1814 when the practice of placing cold water in the bottom of the chamber was superseded by the introduction of a jet of steam: this kept the chamber warm and favoured the formation of a stronger acid. It also opened the way to making the process continuous.

Plans of the St Rollox chambers were supplied to the Newcastle soapmakers Doubleday and Easterby, and on the basis of these the first chamber was erected on Tyneside at the Bill Quay works in 1809.[4] With the subsequent founding of each of the larger alkali concerns, the number of lead chamber sites increased, no doubt being helped by the well-developed lead industry on the Tyne. This growth of sulphuric acid manufacture on Tyneside was to lead to two important technical revolutions. The first related to the source of the sulphur that was burned at the chambers. For a hundred years after the founding of Ward's factory at Twickenham, the sulphur was brought in as brimstone from Sicily: although transport was a burden on this trade, the only serious rivals for sulphur purchase were the wine-makers who used it to dust their vines, and the material could be had in England at £7 per ton.

In 1838 John Tennant of St Rollox and James Muspratt of Liverpool jointly purchased a sulphur mine in Sicily. The Neapolitan Government interpreted this action as a threat to their own profits and placed a duty of £4 on every ton of sulphur exported to Britain. Protests followed (backed up by a show of naval strength—surely the first example of chemical gunboat diplomacy) with the only result that the King of Naples granted a monopoly of sulphur export to a firm in Marseilles, Taix and Co: the price of sulphur in England doubled immediately.

Experiments had already been made on the possible use of pyrites as a substitute for brimstone in the manufacture of sulphuric acid: as early as 1811 a patent had been granted to Thomas Hills and Uriah Haddock of London. The first serious attempt to bypass Sicilian sulphur was made in 1838 by Thomas Farmer, whose vitriol factory at Kennington had been founded in 1778, but there were considerable technical difficulties in the way. Sulphur, once ignited, will burn as long as the supply of air lasts and all the products of its combustion are gaseous. Pyrites, on the other hand, needs to be heated to a high temperature and it leaves behind a bulky residue of metal oxides. In 1840 John Allen of the Heworth Chemical Works in County Durham imported the first cargo of pyrites into the Tyne and for the next fifteen years the large Tyneside alkali makers devoted much ingenuity to the devising of furnaces adequate to the treatment of the new raw material.

The monopoly was broken in 1842, but brimstone never recovered its position as the major source of sulphur. Once installed, the pyrites burners were there to stay, and parallel research into ways of extracting copper from cupreous pyrites made the latter an altogether more attractive raw material. It is nevertheless interesting to record that when a great fire, accompanied by an explosion, occurred in Gateshead in 1854, it was found that very large stocks of brimstone had been stored in a general warehouse. In Lancashire, by contrast, brimstone had quite gone out by 1852. In 1866 Sir Charles Tennant called a meeting in Liverpool of chemical and copper manufacturers to consider how they could obtain control of pyrites supplies. At the end of that year the Tharsis Sulphur and Copper Co. was incorporated, and metal extraction factories were bought up or specially built at Glasgow, on Tyneside, at Widnes, in Staffordshire and at Cardiff. It was a condition of sale to acid makers that burnt ores should be returned to one of the Tharsis factories for copper extraction.[5]

The second important improvement concerned the conservation of nitre. The sole source of this commodity was South America, particularly the coast of Peru. Under the name of 'caliche' it was the mainstay of the prosperity of the country of its origin, and also of a large number of shipping firms engaged exclusively in transporting it to Europe. In the earliest stages of the chamber process all the undecomposed nitre was thrown out with the waste salts in the combustion dish. Later, in conjunction with the external sulphur furnace, nitre pots were built in which the nitre was decomposed with sulphuric acid to yield gaseous oxides of nitrogen: in both cases the oxides of nitrogen were absorbed in the dilute chamber acid and were lost to the atmosphere when the acid was boiled down. In 1827 the French chemist Gay Lussac showed that nitric oxide could be absorbed in concentrated sulphuric acid and constructed a tower at the Saint Gobain works in which this absorption was effected. The problem remained, however, of how to utilize the nitro-sulphuric acid effectively, and in 1859 this was solved by John Glover of Newcastle upon Tyne.

As lead chambers grew in size so the difficulties of making gastight joints of great length increased: consequently a good plumber or lead-burner was assured of steady employment in a vitriol factory. It was in this capacity that Glover entered the chemical industry. Born in 1817, the son of a working man, Glover received only a brief schooling before becoming apprenticed to a master plumber who had the care of the public water troughs in the town: it is said that if the boy was ever held up in his work he would sit in the empty trough to read a book. Hugh Lee Pattinson (the desilverizer of lead) employed Glover at his Felling Chemical Works, first as a plumber, and later in the laboratory where the young man made such progress that Pattinson made him a manager in his new works at Washington, County Durham. Here the first Glover tower was erected in 1859. It was made of firebricks set in molten sulphur, packed with tiles of fireclay and it worked for more than a year—long enough to demonstrate the soundness of the principle of trickling nitrous vitriol down the tower against an upward current of hot burner gases. Two more towers were erected, one at Washington and one at Wallsend where in 1861 Glover founded his own factory; a fourth, at Wallsend, followed in 1864.[6]

The Glover tower was situated between the pyrites burners and the chamber. In this way the hot sulphurous gases collected nitric oxide on their journey up the tower and then passed into the chamber; the de-nitrified concentrated acid flowed into a reservoir at the base of the tower (Pl. 5). The saving of nitre was estimated at 4 per cent of the weight of sulphur burnt. Glover never patented the tower, and indeed showed it to all comers, including the managers of rival factories. Between 1868 and 1870 all the larger alkali works on the Tyne had introduced the tower, which was adopted in Lancashire in 1868 and in London in 1870. In 1871 Georg Lunge published an account of it in *Dingler's Polytechnic Journal* after which it came into general use in Germany.

The earliest method of concentrating weak chamber acid was to boil down in open pans and then to distil in glass retorts, necessarily of limited size. Much thought went into the protection of the glass retorts from the fire over which they were heated, for the breakage of a vessel containing boiling sulphuric acid was obviously a serious matter. Sometimes the retort was put into an iron pot containing sand, or the more ancient method of luting was employed, the bulb of the retort being coated with some specially favoured mixture such as clay and horse dung.

The alternative to glass was platinum, which the misanthropic genius W. H. Wollaston had succeeded in making malleable about 1806. The first large still made by Wollaston's process weighed 423 ounces and appeared in London in 1809.[7] Stills of this kind were often porous and unsound and where oil of vitriol oozed through the metal, repairs had to be made by soldering with gold. After the introduction of Sainte-Claire Deville's method of melting platinum in the oxygen-coal gas blowpipe, this defect was overcome and after about 1860 English platinum stills were mechanically reliable. Chamber sulphuric acid, often made from arsenical pyrites, usually contained small amounts of arsenic and this caused disappointment to manufacturers who thought their acid stills should last for ever.

It has often been stated that the firm of Doubleday and Easterby possessed a platinum still costing £700, and as they were the first

Tyneside producers of sulphuric acid this seems to have been an extraordinarily prescient act of faith. The story seems to stem from an account of chemical manufacture on Tyneside written for the British Association meeting in 1863. There are in Gateshead Public Library large and valuable deposits of inventories and valuations relating to the works on the Durham side of the Tyne, but a search of these has revealed no mention of Doubleday's still.[8] There is, however, a reference to a platinum still and a platinum syphon at the neighbouring works of Lee and Pattinson at Felling, valued at £924 15s in 1840, and it seems likely that this is the still supposed to have been at Bill Quay.

By the 1870s the chamber process had reached a high state of efficiency. The residues from the pyrites burners were sold to iron or copper smelters, spent oxide from gasworks provided a cheap secondary source of sulphur, the nitrous fumes were recycled, control of the ingredient gases ensured continuous production, and the size of the chambers had increased until lengths of 100 feet were by no means rare. It was no longer considered worth the trouble to tinker with a chamber that needed repair; sufficient chambers were usually run to allow the defective one to be dismantled and rebuilt. Yet in what seemed to be its final state of sophistication the process was to undergo further refinement. In 1913 Charles Packard and William Mills patented a new design of chamber, a truncated cone looking somewhat like a sturdy windmill without its sails. These Mills-Packard chambers, further modified by R. T. Maudsley in 1926, had an expectation of life of up to thirty years as against the five to eight years of a conventional chamber.

Contact Process

Even in its heyday, nobody pretended that the chamber process was other than obscure in the details of its chemistry and capricious in its day-to-day working. It is all the more ironical, therefore, that a patent had been granted in 1831 for a process in which the chemistry was apparently simpler and the route more direct. Peregrine Phillips, jun, a partner in a vinegar manufacturing business in Bristol, set out with the utmost clarity his aims and achievements. 'I propose . . . an instantaneous union of the sulphurous acid gas with the oxygen of the atmosphere and thereby save the constant expense of saltpetre, and also the great outlay of capital in the chambers.' This was to be effected by drawing the gases, in correct proportions, by means of an air pump through an ignited tube of platinum or porcelain packed with platinum wire. The sulphur trioxide was to be absorbed in chambers of stone filled with pebbles which were moistened with water or dilute acid.[9]

There were several serious obstacles to the success of Phillips's process in his lifetime. Very little was known about the phenomenon of catalysis and the factors likely to affect a catalytic action adversely; the kinetics of gas reactions had not been explored, and in fact Phillips was wrong in supposing that the gases should be present in the proportion in which they were expected to react; and as important as the lacunae in chemical knowledge was the lack of the necessary chemical engineering know-how to handle gases under critical conditions of temperature and pressure. More than forty years were to elapse before theory and practice were to come together.

The special requirements of the dyestuff trade for concentrated and fuming sulphuric acids caused a number of chemists to seek a better route than the boiling down of weak chamber acid. In 1876 Rudolph Messel and W. S. Squire had some success in combining sulphuric dioxide and oxygen over platinized pumice. Messel's great contribution had been a study of the 'poisoning' of catalysts, especially by arsenic in pyrites. At their Silvertown works Messel and Squire converted chamber acid into fuming acid by subjecting it to decomposition by heat and passing the resulting gases over the platinum catalyst; absorption took place in large Woulfe's bottles (Pl. 7). At the same time Clemens Winkler in Germany was working a similar process, but using platinized asbestos instead of pumice. Both Winkler and Messel followed Phillips in using sulphur dioxide and oxygen in stoicheiometric* proportions—for neither knew the Law of Mass Action, nor had they Le Chatelier's Principle to guide them. The discovery that excess of oxygen assisted the reaction was not made until 1895, and is to be credited to the BASF combine in Germany.

Success in the field of oleum (fuming sulphuric acid) manufacture came at the very time that the decline of the British dyestuff trade (Chapter Nine) caused a slackening in demand. Consequently there was little incentive to manufacturers of acid to convert from chamber

* Stoicheiometry—the fixed proportions in which chemicals combine or react.

to contact process. The outbreak of the 1914 war, with its sudden and imperative call for explosives in addition to dyes, changed the whole picture. At government factories in Scotland, as well as in Lancashire and Yorkshire, contact process plants were set up and in this way the industry was forced to undergo the necessary changes. From that time onward, except for the abnormal period of depression around 1930, the proportion of contact acid in the total United Kingdom output has steadily risen and that of chamber acid has declined until it is now less than one-tenth of the whole. Because contact plants are usually much larger producers than chamber plants the proportion of factories employing chamber processes is nearly a quarter of the total. Platinum catalysts have yielded place to metal oxides, especially vanadium pentoxide, which are cheaper and more resistant to poisoning.[10]

In 1950 geological surveys of the world-famous sulphur deposits in Louisiana seemed to suggest that they were not as extensive as had formerly been thought. Although the fears of early exhaustion were subsequently discovered to be groundless, the immediate cutting down of exports by the American Government highlighted the dependence of British sulphuric acid manufacturers upon overseas sources of their raw materials. The only indigenous supply of sulphur minerals in large quantities lies in the gypsum and anhydrite deposits which occur widely in England but especially in the counties of Durham, Yorkshire and Cumberland. As early as 1903, Georg Lunge, professor of technical chemistry at Zurich but formerly a chemical manufacturer on Tyneside, had suggested the production of sulphur dioxide from anhydrite by roasting it with clay in a kiln: the clinker residue he thought might make cement. This process was attempted in Germany in 1914–18 but without the byproduction of cement.

In 1929 the immense deposits of anhydrite at Billingham were mined and a sulphuric acid plant was brought into operation in 1930: about five years of patient experiment were necessary before the production could be expanded.[11] The anhydrite was mixed with coke or coke breeze and either clay or ash (according to whether or not there was ash for disposal). The mixture was kilned at about 1400°C, the gases scrubbed and dried and, after mixing with the necessary quantity of air, led to catalytic converters. The clinker is ground with gypsum and packed as cement; about one ton of clinker is

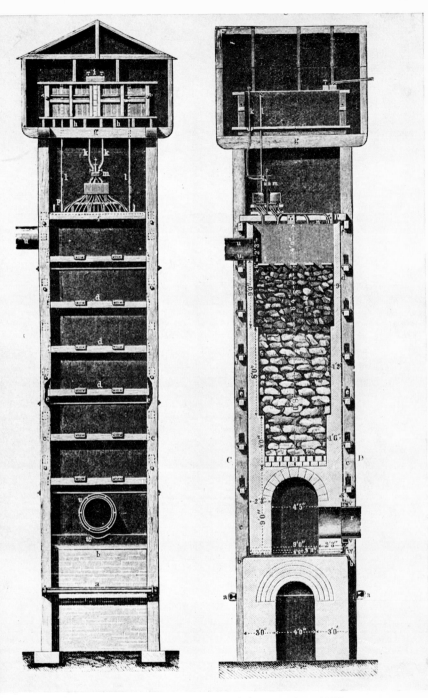

5 Glover Tower for denitrifying sulphuric acid, c. 1870

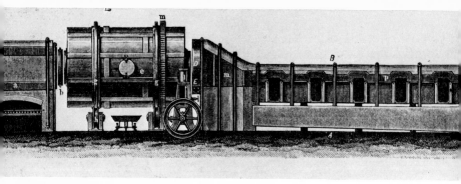

6 Revolving furnace, Williamson and Stevenson, 1870

7 Woulfe's bottles for absorbing acid fumes, c. 1850

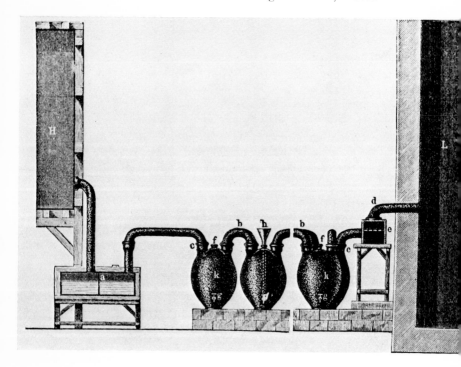

formed for each ton of acid made. The British anhydrite deposits are
unlikely to become exhausted for several centuries. In 1951, under
the threat of the sulphur famine, the United Sulphuric Acid Corporation was founded to work the anhydrite process on information
supplied by ICI. Production began at their works at Widnes in 1954.
In that same year a subsidiary of Albright and Wilson began to make
sulphuric acid at Whitehaven using local anhydrite.

CHAPTER THREE

Leblanc Alkali before 1863

Chaucer's ill-used Canon's Yeoman included alkali in his formidable list of useful chemicals. At that time and for many centuries to come the word denoted a saline product obtained by lixiviating* the ashes of marine plants. So far as chemical technology is concerned the term meant the carbonates of sodium (soda) or potassium (potash). For most purposes these two are interchangeable although there are slight differences in the products obtained; a soap made with potash, for example, is softer than one made with soda. Alkali is an essential ingredient in soap and glassmaking and its supply is therefore of considerable economic concern. By the middle of the eighteenth century three principal sources had emerged, wood ashes, barilla and kelp: there was also a small amount of native soda from Egypt.

The purest form of potash was wood ashes, often recrystallised to form 'pearl ashes'. The main suppliers were Russia and the colonies of America and in both cases transport charges led to high costs of around £25 per ton: by 1802 this had risen to the prohibitive figure of £40. Highly valued by the glass makers was barilla, made from seashore plants found in Spain and southern France and containing about 25 per cent alkali. The barilla trade centred round Alicante and it can easily be seen that political issues would lead to interruption of supplies. The only source of alkali in the British Isles came from kelp, obtained from seaweed and containing a lower proportion of soda in admixture with the potash than was found in barilla. The collecting, drying and burning of seaweed was one of the staple industries of the Highlands and islands of Scotland. Kelp was a variable commodity containing from $1\frac{1}{2}$ to 6 per cent of alkali. The best samples were hard, brittle and resembling indigo in colour; it was, however, frequently heavily adulterated with salt. At a price of £11 per ton it corresponded to alkali at perhaps £300 per ton;

* Lixivation: extraction with water, often called leaching.

nevertheless the demand was such that it was claimed that 100,000 people were employed in kelp making, though the seasonal nature of the trade suggests that many of these would have had some other occupation as well.[1]

Although salts of sodium and potassium are so similar in their chemical and technical behaviour, compounds of sodium are much more plentiful in nature than those of potassium, and the most plentiful of all—sodium chloride or common salt—abounds in sea water and in rock salt deposits. It was natural therefore that chemists should look to common salt as a possible raw material for making soda. Not that the supply of alkali fell seriously short of the demand, but the variable composition of all the natural varieties (except potashes) was a serious technological weakness. Furthermore, the importing of barilla in particular was liable to be cut off by war or government action. These factors contributed to a brisk activity among British chemists and manufacturers relative to the problem of getting soda from salt.

Many makeshifts were resorted to, as glassmakers saved their 'sandiver', the scum from the glass pots, and soapmakers recycled their 'soapers' salts', but there were several ingenious chemical operations in progress. In the years around 1780, for instance, six patents were applied for, all for processes involving conversion of salt to sulphate and subsequent calcination with clay or lime mixed with coal or charcoal. One of the applicants, Richard Shannon of the parish of St Martin-in-the-Fields, claimed that he had spent several thousands of pounds on erecting a factory, John Collison of Battersea had formed an association with a Bristol soaper who was prepared to take his product, and Isaac Cookson of South Shields was eager to supply his own glasshouses with alkali. In the end it was a Frenchman, Nicholas Leblanc, who in 1791 discovered a workable process for converting salt into soda: the British chemists came near to anticipating him and one of them, Bryan Higgins of Greek Street, Soho, came very near indeed.[2]

Leblanc Process

Table 3 shows a flowsheet of the Leblanc Process in its earliest form. Salt was decomposed with sulphuric acid with the formation of sodium sulphate (saltcake) and the consequent liberation of hydrochloric acid gas. In this operation the semi-liquid mass was heated in

TABLE 3
FLOWSHEET OF LEBLANC PROCESS IN EARLY FORM

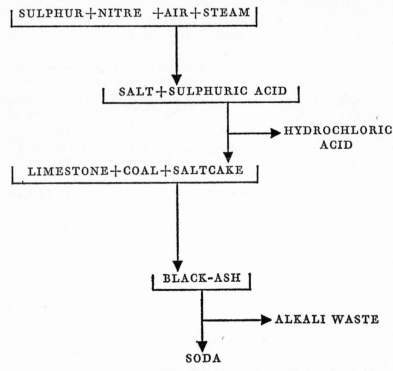

a reverberatory furnace with periodic raking to assist the escape of the hydrochloric acid gas. (Fig. 3). This section of a works was always marked by the presence of a tall chimney which drew the fire and helped to dissipate the acid fumes into the upper air. The hot mixture of sulphuric acid and salt was extremely corrosive, so that the iron pans and brick linings of the furnaces needed frequent renewal, as did the implements used for raking: we shall discuss the effect upon the operators later (p. 48). The chemical reactions took place in two stages, denoted by the equations

$$NaCl + H_2SO_4 \longrightarrow NaHSO_4 + HCl$$
$$NaCl + NaHSO_4 \longrightarrow Na_2SO_4 + HCl$$
$$\text{(saltcake)}$$

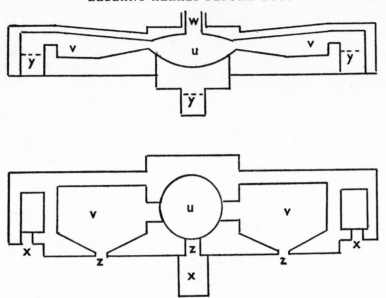

Figure 3 Section and Plan of Saltcake Furance

u. Saltcake pot
v. Roaster bed
w. Flue to acid condenser
x. Fire doors
y. Fire grates
z. Working doors

The saltcake was trundled in barrows to the black-ash section where it was heated with coal and limestone: at first this was carried out in furnaces similar to those used in making saltcake, but later the famous 'revolvers' (mechanical furnaces for mixing the ingredients while they were being heated) were introduced (Pl. 6). The black-ash was run into bogies, square iron boxes on wheels, and then allowed to cool. The cold lumps of cuboid form were by some curious logic termed 'balls', and the building in which they were collected was known as the 'ball-house'. The reaction was complicated in its mechanism but the overall book-keeping can be represented:

$$Na_2SO_4 + 4C + CaCO_3 \longrightarrow Na_2CO_3 + CaS + 4CO$$

The crude ball soda was extracted with water in lixiviating tanks; hence the soda passed into solution and the calcium sulphide and

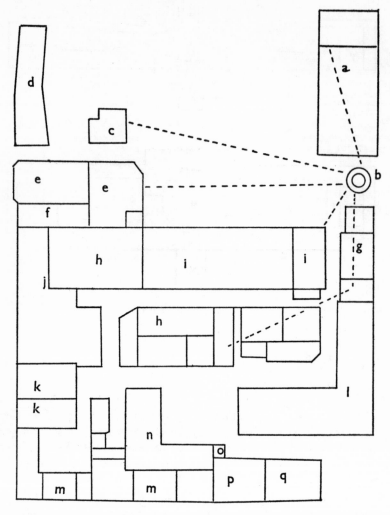

Figure 4 Plan of Heworth Chemical Works near Newcastle, 1836
An unusually compact layout, not originally designed as a Leblanc soda
works.

a. Sulphuric acid chambers and burners
b. Chimney. Dotted lines show underground flues
c. Lime kiln
d. Sulphate warehouse
e. Acid condensing yards

f. Saltcake furnaces
g. Workshops
h. Carbonation
i. Boiling and crystallising
j. Filtering sheds
k. Epsom salt drying
l. Blackash and lixiviation
m. Dwelling houses
n. Magnesia sheds
o. Dog kennel
p. Stables
q. House with lawns and shrubs 'for a respectable family'

coal ash were left behind. Waste heat from one or other of the furnaces was employed in evaporating the soda solution (tank-liquor) to give the product 'soda ash' or anhydrous sodium carbonate. The solid residue (tank waste) was either tipped on a heap outside the factory or taken out to sea in barges.

The outward form of an early Leblanc soda works would thus present itself as a collection of long low buildings to house the furnaces, overlooked by the lofty rectangular pile of the acid chambers and dominated by the tall chimney from which clouds of acrid fumes would issue at intervals of a few hours. Outside the walls, and gradually attaining an eminence of its own, would be the waste heap, a dark grey running eyesore at all times and in wet weather the source of a foetid sulphide smell.

The first works in England to operate the Leblanc process was at Walker, a few miles east of Newcastle upon Tyne, and was owned by the brothers John and William Losh. John Losh of Woodside near Carlisle had received some training in chemistry at Cambridge in Bishop Watson's time. William had also studied chemistry at Cambridge and had gained metallurgical experience in Sweden before working with Lavoisier in Paris. He had watched the progress of French experiments in soda making, staying on in Paris until it was no longer safe for an Englishman to do so. In cooperation with Archibald Cochrane, ninth Earl of Dundonald (a man of wide scientific sympathies, broad technical imagination, but no business sense), the Losh brothers had begun experiments on soda production in 1793. Four years later there passed into their hands a share in Walker Colliery, where ten years earlier the King Pit had been sunk and had been fitted with a Boulton and Watt engine on the 'sun and

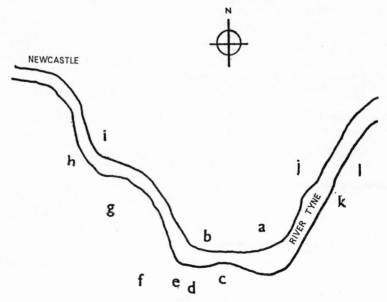

Figure 5 Site of Losh's, the first Leblanc Soda Works in England, showing adjacent chemical development

a. Losh's Walker Alkali Works
b. c. Cook Brother Alkali Co
d. Burnett & Sons Chemical Co
e. Imeary's Alkali Works
f. Pattinson's Felling Chemical Works
g. Friars Goose Chemical Works
h. Allhusen's Newcastle Chemical Co
i. Langdale's Chemical Manure Works
j. Hunter's Chemical Works
k. Tharsis Sulphur & Copper Co
l. Tennant's Hebburn Alkali Works

planets' principle. The pit had struck a brine spring which very much hindered the production of coal: the strength of the brine was about five ounces of salt to the gallon. Since brine was the ultimate starting point in all soda making enterprise, Losh and Dundonald were prompt to sign a lease with Newcastle Corporation for the sole use of the brine, and another with the owners of the colliery to allow the engine to pump the brine to an adjacent works.[3]

To have recovered the salt by simple evaporation would have in-

curred a duty of £36 per ton. Concessions were available to soda
makers, provided that the salt was spoiled for domestic use, and in
October 1798 an Act of Parliament setting out new conditions for
the imposition of the salt tax included a paragraph granting exemption to the Walker users. The Walker Works thereby acquired an
advantage over competitors, but there is no reason to believe that
similar concessions would not have been granted to others who cared
to ask.

At first the Walker Works produced soda on a small scale by a
number of processes, but chiefly by one discovered by the Swedish
chemist Scheele. In this, salt and lead oxide (litharge) were stirred
into a paste with water and left for a few days. Caustic soda was
formed according to the equation

$$NaCl + H_2O + PbO \longrightarrow NaOH + Pb(OH)Cl$$

and was carbonated by burning with sawdust; the basic lead chloride
was sold as a yellow pigment under the name of Turner's Yellow.
Russian potashes were converted to soda by double decomposition
with salt, and whenever the price made it worth while soapers'
waste, sandiver from the glass pots, and nitre cake from the nitric
acid manufacturers were all worked up into soda. The company also
made litharge (for the Scheele process), roll and flour sulphur,
charcoal and pyroligneous acid from wood, and for a time they engaged in silver refining. Between 1802 and 1806 the Leblanc process was introduced (Losh having paid further visits to Paris), and
soon it displaced the other soda processes: by 1823 the works had
become solely a Leblanc soda factory.

The next Leblanc soda factory in England was that of James Muspratt, founded in Liverpool in 1823. Muspratt was born in Dublin in
1793, and like many others who made their mark in chemistry was
apprenticed to a wholesale druggist and apothecary. There followed
periods as a soldier of fortune in the Spanish peninsula and as a naval
officer (from which post he deserted), but in 1819 he set up in Dublin
as a manufacturer of acetic acid, turpentine, hydrochloric acid and
later of prussiates of potash. His partners in this small chemical
venture would not back him in making soda so he came to Liverpool alone, leasing land on the canal bank in Vauxhall Road. Being
without capital to build the necessary lead chambers, he was forced
at first to continue in the manufacture of prussiate until sufficient

profits had accrued to enable him to proceed with all the business of a Leblanc alkali works.[4] It was characteristic of Muspratt that both history and geography should favour him: the time was appropriate for 1823 saw the removal of the salt tax, and the place presented the advantages of a port in close proximity to a salt district and to a coalfield.

At first the local soapers showed a deep prejudice against synthetic soda, having learned empirically how to make use of only natural potash. It is said that Muspratt was forced to give away free consignments of soda, and to superintend the soap-making process in neighbouring works in order to prove the worth of his product. But prove it he did, and to such a degree that the Liverpool site had soon become too restricted to allow further expansion. In 1829, therefore, James Muspratt joined J. C. Gamble in building a factory at St Helens and two years later another was built at Newton-le-Willows (this time without Gamble). Soda-making in the 1830s was largely a rule-of-thumb business, but Muspratt had the foresight to engage James Young (the founder of the Scottish paraffin industry) to superintend the laboratory at the Newton works. Persecution by landowners and by Liverpool Corporation on account of the blighting effect of acid fumes caused Muspratt to close down the Liverpool and Newton works in 1849. In 1850 works were started at Flint and at Widnes; these passed into the hands of Muspratt's sons and eventually were swallowed up by the United Alkali Co., in 1890.

Ten years after Muspratt's entry into the field of alkali manufacture there was founded on Tyneside a venture from which several important developments were to spring. This was the works known variously as John Lee and Company and as Felling Chemical Works, in which the most notable partner was Hugh Lee Pattinson. Pattinson was born in Alston, Northumberland, in 1796. After a meagre education in the village school and a period of service in his father's drapery business, he came to Newcastle as a clerk at Anthony Clapham's soapworks. Here he acquired sufficient skill in chemistry to be appointed in 1825 Assayer of the Alston Moor Lead Mines on behalf of the Commissioners of Greenwich Hospital. While engaged in that thankless work he brought out his famous process for the desilverization of lead, and out of the proceeds from this he joined in the establishment of the Felling Chemical Works (Pl. 9).[5] Originally covering eleven acres, the works grew rapidly until well over seventeen acres were occupied by 1848. The *Penny Magazine* for 1844 carried an article about this works as part of a series describing typical factories. From this we learn that in addition to the usual furnaces, chambers, towers and sheds, the building included a cooperage where the barrels for soda were made, a sawpit, a joiners' shop, a blacksmith's and a plumber's shop: there was also a works laboratory (not by any means a universal feature in the 1840s). The plumber's shop and laboratory are of particular interest for it was through these that John Glover entered the chemical industry.

The decomposing house, in which salt was heated with sulphuric acid to form saltcake, measured 240 feet by 44 feet and contained six furnaces. To convey the large quantities of salt from the river frontage to the decomposing house, and also to bring saltcake to the black-ash department, an overhead railway was constructed at a height of 25 feet. Such elevated railways (known as 'gears') seem to have been popular with Leblanc soda manufacturers. It was an unusual feature of Pattinson's Works that no elaborate precautions were taken against the leakage of process information: in 1838 the members of the British Association for the Advancement of Science, meeting in Newcastle, were invited to inspect the Works and perhaps this is why the publishers of the *Penny Magazine* selected the Felling Works to illustrate alkali manufacture.

In 1842 Pattinson founded the Washington Chemical Company. This works, on the County Durham site from which the family of President George Washington came, is now part of the Turner Newall group. Until 1970 magnesia was still made there by the process of extraction from local dolomite which Pattinson discovered; it was here also that the first Glover tower was erected. Pattinson died in 1858, but the Felling Works continued in operation until 1886; when it closed 1,400 men were put out of work. Pattinson's son-in-law, R. Stirling Newall, joined the firm in the 1860s.

Of the other Tyneside factories, about two dozen in all, only one calls for notice at this point. This was the world-famed Newcastle Chemical Company founded by Christian Allhusen in 1840 and surviving until the formation of Imperial Chemical Industries Ltd in 1926. Christian Allhusen had no chemical training. Born in Kiel in 1806, he came to Newcastle in 1825 to engage in business as a corn merchant.[6] When he died in 1890 his will was proved for £1½ million, most of which had been amassed on the banks of the Tyne.

Tynesiders, however, never learned even to pronounce his name correctly, the factory being referred to as 'Alley-Hewson's'. All-husen's business interests extended to shipping (the firm later known as Borries Craig and Co.), steelmaking (Bolckow and Vaughan), water-works, railway and banking, and it was in this spirit that he bought up an unproductive glass, soda and soap works and made it into one of the world's largest chemical factories.

From the start the venture enjoyed a number of advantages. The site was surrounded by land that was already industrialized and so the risk of being indicted as a nuisance was reduced. The river frontage was almost opposite to Newcastle quayside, and therefore at the centre of the coal trade and well placed for distributing the firm's products. The most important factor in the firm's success, however, was undoubtedly the Allhusen policy of putting to work any new process or plant that was seen to be effective. Thus, although none of the dramatic discoveries or inventions that altered the shape of the alkali trade was ever made at the Gateshead works, the works was constantly being brought right up to date. By far the largest section of the works was that devoted to the manufacture of sulphuric acid. Thirty chambers were accommodated in four groups, each of approximately 200 feet square. These groups of chambers were covered in wood and the tops were fitted with turrets and battlements giving the appearance of castles. As expansion gathered speed, several sections were duplicated on an adjacent site, resulting in a 'high works' and 'low works' system. Connections between the two centres were maintained by means of the familiar elevated railway. To this day, when an elderly man in Gateshead or Newcastle says that he 'worked at the chemicals' it is almost always Allhusen's that is meant[7] (Pl. 11).

The remaining important alkali works are best considered in Chapters Four, Five or Six. It is sufficient at present to note that the Leblanc process was beset by a number of difficulties inherent in its nature. First, it was extremely wasteful since the chlorine went into the atmosphere and the sulphur went onto the waste heap. Moreover, the nature of the central reaction, the conversion of saltcake to black-ash, was obscure and amenable only to empirical control. The carbon monoxide liberated in the reaction would catch fire at the surface of the semi-liquid mass and burn with a golden sodium flame: this phenomenon of 'soda candles' was used to judge the progress of the reaction. If heating was discontinued too soon after the disappearance of the candles, the product was known as 'green balls' and contained unchanged saltcake; if on the other hand the mixture was heated too long 'burnt balls' resulted and too much sodium sulphide was formed.[8] A careful and experienced worker could estimate the most favourable moment for pouring the melt, and upon his judgment the success of the factory depended in no small degree. A further weakness in the structure of the trade lay in the boom-town setting of the alkali districts; by contrast with the robust building of, say, a cotton mill, a soda works was hastily thrown together out of sheds of the flimsiest construction and brickwork often only one course in thickness. Small wonder that an alkali maker asked during a period of recession why he did not shut down his works for a time, replied that if he shut the plant down it would fall down.

CHAPTER FOUR

Leblanc Alkali after 1863

We have seen in Chapter Three how all the chlorine in common salt escaped into the atmosphere as hydrochloric acid gas, but the scale of the loss (and the parallel threat to the amenities of an alkali district) becomes apparent when the stoicheiometry of the reaction is considered. Every ton of salt, in the course of its conversion to saltcake, yields $12\frac{1}{2}$ cwt of this choking, corroding gas, and since a typical alkali district such as Tyneside or Merseyside might easily be decomposing 100,000 tons of salt in a year, something more than 60,000 tons of hydrochloric acid gas would pour out over the countryside. As early as 1839 a petition relating to Losh's factory (the first Leblanc works to operate in England) complained that

> the gas from these manufactories is of such a deleterious nature as to blight everything within its influence, and is alike baneful to health and property. The herbage of the fields in their vicinity is scorched, the gardens neither yield fruit nor vegetables; many flourishing trees have lately become rotten naked sticks. Cattle and poultry droop and pine away. It tarnishes the furniture in our houses, and when we are exposed to it, which is of frequent occurrence, we are afflicted with coughs and pains in the head . . . all of which we attribute to the Alkali works.[1]

Hugh Lee Pattinson's factory was under regular observation by a local land owner who kept in a diary precise records of dates and times when the noxious white cloud of acid gas rolled forth over his land. One of the tenants who suffered a good deal of damage kept some dairy cows, but he was unwilling to testify against the factory as many of the workers bought milk from him. This episode highlights the very complex nature of the pollution problem: on the one hand trade and employment were brought into a district, whilst on the other the amenities of the district were destroyed. Such a division of interests occurred at the South Shields factory of Cookson's (alkali, glass and lead manufacturers, ultimately absorbed into the Associated Lead subsidiary of Goodlass Wall and Co. Ltd). In 1843 when the firm was going through a particularly bad period of harassment for damage to crops and pasture, the workmen held a grand procession to the Town Hall to enlist public sympathy for their employers. One of their number, a man of immense stature and healthy complexion, allowed himself to be exhibited as a living proof of the beneficial effects of the atmosphere within the works.[2]

The manufacturers themselves fought back against the rising tide of complaint and litigation. In *Chemical News* for 1860 a letter headed 'Is hydrochloric acid a cure for consumption?' was signed by one who called himself 'An old alkali manufacturer'. Part of the letter must be allowed to speak for itself.

> The foreman of the upper works had a lodger, a young man, by trade a ship carpenter, who was very ill and much reduced, evidently in a consumption. One fine day he asked leave to go up in the barge to the upper works, where he remained while the barge was being loaded, all the time breathing an atmosphere highly charged with hydrochloric acid gas. He experienced great relief from this, and often repeated his visit. He regained his strength rapidly, so that he was soon able to walk to the works, so eventually I gave him employment there and he became a strong robust workman.

At Jarrow, too, the chairman at a counter-protest meeting said, 'Gentlemen, some people say the manufactories of this borough are injurious to health. I don't believe it: The healthy faces of everyone around me prove it false. Look at ourselves—Why, this is one of the healthiest towns in all Her Majesty's dominions.'

We have already alluded to the troubles which led Muspratt to abandon his Liverpool and Newton works, and start up the Widnes factory which was to keep his name alive for many decades. Of his St Helen's factory it was said:

> They invaded a very prettily situated, nice little country town. It was the residence of several well-to-do families, who lived in substantial comfortable homes attached to their shops, or close to their business. Gardens and well-stocked orchards ran from street

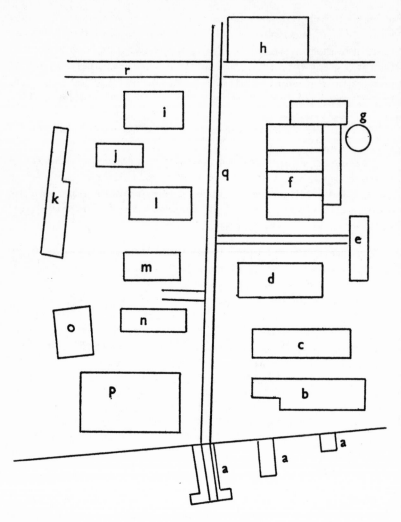

Figure 6 Layout of Curlew Chemical Works, Jarrow on Tyne, 1875. This late design shows a works intended mainly to supply bleaching powder for an associated paper works.

a. Jetties
b. Lixiviation and carbonation
c. Blackash furnaces
d. Saltcake furnaces
e. Pyrites burners

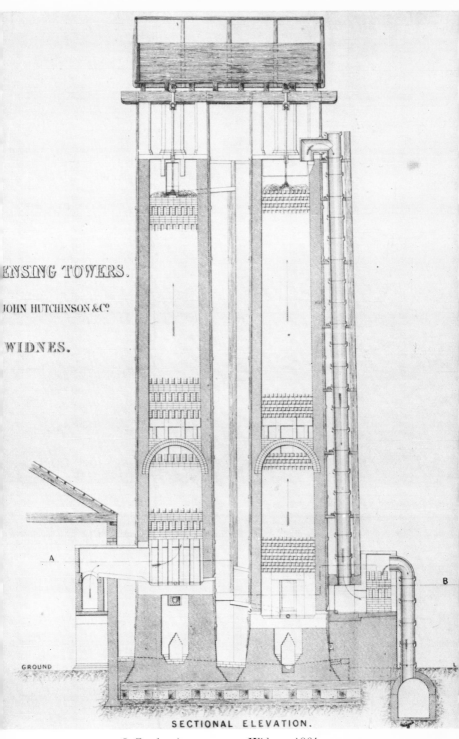

ENSING TOWERS.

JOHN HUTCHINSON & Cº

WIDNES.

A

B

C

GROUND

SECTIONAL ELEVATION.

8 Condensing towers at Widnes, 1864

9 Sulphuric acid
chambers at Fel[...]
1843

10 Flues for coo[...]
acid gases before
absorption, Run[...]
Alkali Works, 18[...]

f. Vitriol section, comprising 5 chambers with concentrating and
    denitrifying towers
g. Chimney
h. Esparto grass store
i. Office and laboratory
j. Sawmill
k. Workshops and stables
l. Hydrochloric acid condensers
m. Salt house
n. Lime kilns
o. Weldon chlorine plant
p. 5 bleaching powder chambers
q. Overhead railway with branches
r. Public road

to street, the roads that led out from the town were lined with
avenues of trees, and on all sides were rich farm lands, with well-
cultivated hedges and abundant timber. The streams . . . were
stocked with trout. If, when these founders of the industry that
would so enrich this town could, when they were selecting the site,
have had a vision of the transformation they would initiate, they
might have shrunk from the enterprise.

Such was the view, not of a residents' association, but of the
Chemical Trades Journal in 1889.

The same *Journal* makes nostalgic reference to the pre-Leblanc
state of Widnes:

Fields with green hedgerows and healthy trees still extended
along the riverside past the old 'Snig-Pie House'. This noted
hostelry was still a favourite resort for picnic parties. The Widnes
Marsh was used by the neighbouring farmers to graze their
cattle on.

By contrast the *Victoria County History* could state:

A district more lacking in attractive natural features it would be
difficult to conceive. A great cloud of smoke hangs continually over
the town, and choking fumes assail the nose from various works.
In the face of such an atmosphere it is not to be wondered at that
trees and other green things refuse to grow.

It is obvious that legislation would have to be introduced to curb
the devastation of the countryside, and in 1863 the Alkali Works Act
(26.27 Vict. 124) was passed. By the terms of this Act, manufacturers

had to condense at least 95 per cent of the hydrochloric acid gas, and an Alkali Inspectorate was set up under Dr Angus Smith to see that the law was obeyed. The story of how manufacturers coped with the problems of this and other waste products is very largely the story of the alkali trade for the remainder of the nineteenth century, for in one way and another the Leblanc method for making soda developed from the wasteful crudity of its beginnings into an elegant and sophisticated process.

To be fair to the alkali makers, many had attempted the condensation of hydrochloric acid but had given it up as a hopeless task. Series of Woulfe's bottles (Pl. 7) had been used, as had large earthenware cylinders, but these had been quite inadequate to deal with the rapid gush of gas which resulted from the initial mixing of salt with sulphuric acid. In several works lengthy underground flues had been constructed on the lines of those that had proved successful in lead smelting; these were effective in cooling the gas but did not solve the main problem which was to hold the gas in contact with water long enough for solution to take place.[3] Both Pattinson and Allhusen spent thousands of pounds in this way but to little avail: in 1839 Pattinson wrote to a complaining landowner: 'We had expected to give you no cause for complaint this summer, and we have been at considerable pains and expense in putting up condensing apparatus, in which we are prepared to show we have been completely successful.' Muspratt on the other hand told William Gossage that 'all the waters of Ballyshannon itself, would not suffice to condense the acid I make'. (See Pls 8 and 10).

If the acid could not be contained within the works, the next best expedient was to disperse the gas over as wide an area as possible in order that its effect might be spread thinly instead of concentrated on one or two fields or gardens. This hope called into being the tall chimneys which were so salient a feature of the alkali works. Perhaps the most famous of these was 'Tennants Stalk' at the St Rollox Works in Glasgow. The chimney, founded on a bed of solid sandstone twenty feet below the surface of the ground, was fifty feet in diameter at its foundation, forty feet at ground level tapering to fourteen feet six inches at the top; the height was 420 feet. 'The chimney when completed will be elevated upwards of 600 feet above high water level at the Broomielaw; and will be an object of magnificent simplicity, and present to the traveller a landmark of colossal dimensions. The stalk . . . will be about the height of the great Pyramid of Egypt.' For a few weeks after its completion in 1842 the chimney was not connected to the plant but was open to picnic parties who were conveyed to the top to enjoy the view over the surrounding country: the chimney survived for eighty years.[4]

Anthony Clapham's chimney at Friar's Goose was also the scene of jollification. It was built nine years before Tennant's and was much smaller, but nevertheless an impressive structure of 263 feet in height tapering from 27 feet to 7 feet at the top: local pride exulted in its being 38 feet taller than Muspratt's chimney at Liverpool. A local paper reported: 'Mr Clapham . . . entertained a party of friends with a sumptuous repast at the bottom of the chimney, to the great delight of his friends, who expressed their surprise and astonishment at this stupendous work of art.' Several alkali works chimneys were struck by lightning, some blew down in gales, and some collapsed due to weaknesses in construction. So far from solving the pollution problem, they helped the accusers of the alkali makers to identify with certainty the particular factory from which hydrochloric acid was emanating at a specific time; the records of petitions abound in such phrases as 'June 24th, at 10 a.m. a large white cloud was seen to emerge from Mr Lee's chimney'.

The device by which the hydrochloric acid was eventually trapped, under threats of the 1863 Act, had been invented in 1836. William Gossage (Chapter Six) put into operation the tower which bore his name at the alkali works which he ran at Stoke Prior, in the Worcestershire brine spring area.[5] The principle of the Gossage tower consisted in passing the gas upwards through a deep bed of coke down which a slow stream of water flowed, the counter-current process being aided by the enormous surface area which the coke presented. The towers were interpolated between the saltcake furnaces and the chimneys which now fulfilled the more reputable function of providing sufficient draught to pull the heavy gas through the towers. The alkali manufacturers now found themselves in possession of large quantities of unsought hydrochloric acid for which there was little sale, and it is said that many—having ceased to pollute the air—now polluted the rivers instead.

In almost all the larger works the hydrochloric acid was made into chlorine for bleaching powder. The story of this chemical is told in Chapter Six; our purpose here is to consider the troubles which

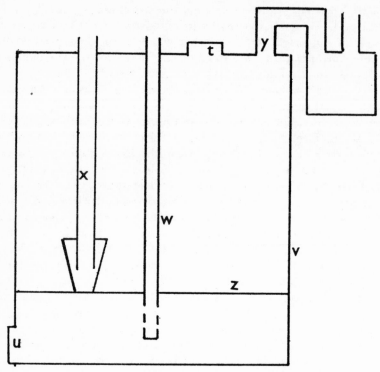

Figure 7 'Still' for making Chlorine from waste Hydrochloric Acid

t. Filler hatch for manganese dioxide
u. Outlet for manganese residues
v. Vessel made of Yorkshire flagstones
w. Steam inlet for heating
x. Inlet for hydrochloric acid, with liquid seal
y. Chlorine outlet with trap
z. Grid supporting lumps of manganese dioxide

the alkali maker underwent, and the consequent changes in the organization of a chemical works, in pursuit of an economical way of affecting the oxidation of hydrochloric acid to chlorine. The oldest chlorine process was that by which the gas had been discovered by Scheele in 1770, involving manganese dioxide in the form of the mineral pyrolusite:

$$MnO_2 + 4HCl \rightarrow MnCl_2 + 2H_2O + Cl_2$$

The equation reveals that half the chlorine and all the manganese was lost. In practice the loss was greater than that predicted by the stoicheiometry, and was estimated at 16 cwt of pyrolusite for every ton of bleaching powder.

A successful method of recovering the lost manganese was evolved between 1866 and 1869 by Walter Weldon of St Helens, on the basis of experiments which he carried out at the Walker Chemical Works.[6] Weldon was not a chemist by profession but a journalist and publisher: he founded and edited a number of household and women's journals including a famous range of patterns for home dressmakers. Milk of lime (a suspension of lime in water) was mixed with the waste manganese liquors and a current of air was blown through the mixture. Manganese dioxide in a form known as 'Weldon Mud' was regenerated as shown by the equations:

$$MnCl_2 + Ca(OH)_2 \rightarrow Mn(OH)_2 + CaCl_2$$
$$Mn(OH)_2 + \tfrac{1}{2}O_2 \rightarrow MnO_2 + H_2O$$

The Weldon Mud was pumped back to the chlorine stills (Fig. 7) and used instead of fresh pyrolusite. Some of the smaller firms engaged in the alkali trade could not afford the capital outlay for the necessary plant but the more progressive firms devoted whole sections of their sites to 'the Weldons'; the noise of the air blowers was a feature of these parts of the works.

A second process for making chlorine from hydrochloric acid gas without trapping it in towers and without the use of manganese, was worked out by Henry Deacon between 1868 and 1870.[7] Deacon's background was very different from that of Weldon. Born in London in 1822 Deacon enjoyed the family friendship of Michael Faraday. He gained engineering and foundry experience with James Nasmyth (of steam hammer fame), received managerial training at Pilkington's glass works at St Helens and served under John Hutchinson at Widnes before founding the firm which came to be known as Gaskell, Deacon and Co. In Deacon's process, hydrochloric acid gas from the saltcake furnaces was mixed with four times its volume of air and passed over clay balls dipped in a solution of copper chloride; these 'Deacon marbles' were later replaced by layers of broken brick, though the marbles continued in general use on the Tyne. The chemical reaction was not new, but Deacon had both the persistence and the engineering knowledge to translate a piece of laboratory apparatus

into an industrial plant. Among the difficulties that he had to overcome was that of balancing on the well-known chemical tight-rope,
where increased temperatures on the one hand speed up the reaction
but on the other reduce the proportion of reactant converted. Deacon
chlorine was diluted with nitrogen and a special kind of bleaching
powder chamber was required to absorb it.

The flowsheet for the original Leblanc process shows that all the
sulphur from the sulphuric acid found its way onto the waste heaps
which accumulated near to the factories. Apart from the nuisance
created by the smell from the waste heaps (calcium sulphide in an
atmosphere rich in acid fumes!) the loss of sulphur was a severe
economic drain on the process. Gossage had tackled the problem in
1837 and several Tyneside firms had recovered some of their sulphur
by partial oxidation to thiosulphate. This was sold to paper mills for
use as an 'anti-chlor' in removing the last traces of chlorine from
their bleached raw materials.

Ludwig Mond brought his genius to bear on the problem, using
waste hydrochloric acid to precipitate sulphur from the thiosulphates.
His concern for the grave losses incurred in dumping sulphur wastes
expressed in a speech in 1869, would be shared by many of his
hearers.

> I see here several gentlemen who have spent thousands of pounds
> in trying to turn this obstinate substance (alkali waste) to advan-
> tage. I see many more still who spend many hundreds each year
> to have it carried away to the dumb fishes of the sea who cannot
> petition Local Boards or Parliament, nor bring a suit for nuisance,
> and probably there is nobody in this room who has not more than
> once had his nose offended and his appetite spoiled by coming into
> too close a contact with the waste heaps in this neighbourhood.

Mond's process was carried out at the Widnes factory of John
Hutchinson and subsequently adopted by Charles Tennant at
Glasgow and by Muspratt's works at Widnes: the initial oxidation
was slow, however, and none of the expedients adopted to speed up
that part of the scheme was really effective.[8]

The only really satisfactory process for recovering sulphur from
alkali waste was announced by Alexander M. Chance in 1882 and
combined with that of C. F. Claus in 1888.[9] Carbon dioxide was blown
through a slurry of alkali waste in water to liberate hydrogen sulphide

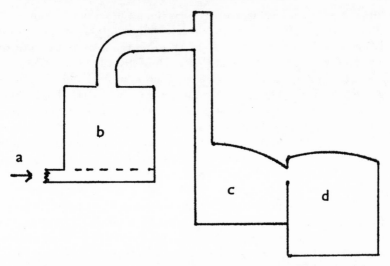

Figure 8 Claus Kiln for Sulphur Recovery as used at Chance's Oldbury
Works in 1887

a. Inlet for hydrogen sulphide and air
b. Iron oxide chamber
c. Brimstone chamber
d. Flowers of sulphur

which was burnt with the calculated quantity of air in a Claus Kiln.
Fig. 8)

$$CaS + H_2O + CO_2 \rightarrow Ca\,CO_3 + H_2S$$
$$H_2S + \tfrac{1}{2}O_2 \rightarrow H_2O + S$$

The above reactions represent only the overall book-keeping, and
are not to be taken as indicating reaction mechanisms. The sulphur
from this process is extremely pure.

It was the application of Claus's patent of 1883 that tipped the
balance in favour of Chance's process. The saving to the alkali manufacturer of sulphur recovery by the earlier processes was of the same
order as fluctuations in the price of pyrites: accordingly, when the
Tharsis Sulphur and Copper Company reduced its selling price in
1883 it was no longer worth working the old sulphur plants, and
Chance's own plant at Oldbury stood idle. No wonder that Eustace
Carey described sulphur recovery as a 'sort of will-o-the-wisp in the

alkali trade that had lured many a good man to serious discomfiture if not to ruin'. Claus's Kiln (in which hydrogen sulphide and air were passed over iron oxide maintained at a dull red heat by the heat of reaction) was being worked at the South Metropolitan Gas Works in London. Chance obtained samples of sulphur from this source in order to compare them with the product from his own Oldbury plant. He confessed that, though the problem seemed simple on paper, four years of labour and the further expenditure of several thousands of pounds were necessary before he was able to make pure sulphur from alkali waste on a manufacturing scale and at an economical cost. He also expressed his conviction that if Gossage had had at his disposal in 1837 the machinery and appliances and particularly the powerful carbon dioxide pumps that he himself had fifty years later, Gossage would easily have hit upon the same process.

With the achievement of this last success in waste product utilization, the whole character of the alkali works was changing. The principles of chemical engineering were replacing the wheelbarrow and shovel methods of the older sections of the works, and the instinctive judgment of the experienced chargehand was giving way to precise control by the chemist. At the same time the chemical economy of a soda works was becoming uncomfortably complex. In 1840 a works might have taken in sulphur, nitre, salt, coal and limestone and made only soda as a saleable product. By 1890 the same works would take in pyrites, salt, nitre, coal, limestone and pyrolusite, making products that would include soda, caustic, bicarbonate, hypo, bleach, several grades of sulphuric acid, hydrochloric acid, chlorates and hypochlorites, sulphur, sulphides, calcium sulphate and Epsom salt. In addition the iron and copper residues from the pyrites would be worked up either in the alkali works or in some adjacent factory. Since there was some degree of stoicheiometric relationship between the quantities of products, the Leblanc process could easily become unbalanced: if the demand for one product increased while that for another decreased, some part of the system had to be run at a loss. In fact soda, the very product which called the heavy chemical industry into being, was often made at a loss so that bleach could be produced at a profit.

Legislation, too, was beginning to weigh heavily on the smaller or old-fashioned business in which the margin between success and failure might be dangerously narrow. The Factory Acts, from 1878 to 1895, began to change the layout of chemical works. Soda vats were to be covered, gangways and gang-planks were to be fenced, dangerous places illuminated and where Weldon chlorine was used tests had to be made on the chlorine content of the atmosphere and the figures recorded in a test book. Respirators had to be provided for rescue purposes and places set aside for washing down those splashed with acids or with caustic.[10]

The workers also were changing with the changed conditions in the works. The labour force was originally made up of Irish immigrants, some of whom had left home to become canal-diggers (navigators or 'navvies'). To these men the alkali trade offered long arduous work, often of a dangerous nature, but without the necessity to keep to the rhythm of a fast machine, as in a textile mill. The disposition of plants was unique, the very open aspect providing the opportunity to walk about (and perhaps go missing for an hour or two). It was natural that men exposed to heat and choking gases should be heavy drinkers and there was always a public house nearby, or even within the boundary of the works; the drink—beer and whisky—was brought to the plant by a boy. The effects were described by one who had compared conditions at Widnes and Newcastle and found them identical.

> The beginning of the week is, of course, the worst. On Mondays the foremen are chiefly employed in keeping their men in an upright position, and not until Tuesday or Wednesday is the maximum angle of stability generally secured. . . . During two or three days of the week the bulk of them would furnish happy studies for an artist of the Cruikshank school—but I would not like to be the artist.

Even if we allow for some prejudice in that account, it remained true that life in a chemical works unfitted a man for any other kind of job.[11]

The workers were divided into process men, who actually made the chemicals, and yard men who trundled barrows from place to place. Yard men whose health was too far gone for heavier work were put to breaking up large pieces of pyrites into a size suitable for the burners; this was a sitting down job. The process men belonged to one of the five groups into which the works was organized, vitriol, saltcake, revolver, finishing or bleach. In the vitriol section most men were pyrites burners. The saltcake men (known as 'pot-men') were the roughest men in the works; their job was to charge, rake and empty the saltcake furnaces, in the course of which they were exposed to gushes of hydrochloric acid gas. R. H. Sherard, in *The White Slaves of England*, wrote that a saltcake man could be recognized by his toothless mouth and that the saltcake department (decomposing house or pot-house) was characterized by the heaps of bread crusts which the men could not eat.

The revolver men made the black-ash, and theirs was a responsible job: upon the senior workmen in this section rested the decision of when to pour the mass (p. 35). Finishing men worked on the lixivia-tion and crystallization of soda and sometimes on the treatment with lime to yield caustic soda. It was not a particularly arduous job, but it carried the risk of falling into a vat. The elite of the works, however, were the bleaching powder men, bleach packers and lime dressers. The bleaching powder chamber is described in Chapter Six. The lime dresser spread the lime on the floor or on shelves, an operation which was not dangerous but merely unpleasant; his hands and arms were smeared with grease, his trouser-legs wrapped with sheets of brown paper and, theoretically, his face protected by a mask and goggles. The bleach packer raked out the chlorinated lime into casks. He was similarly protected, for he had the additional hazard of chlorine gas to contend with (Pl. 16). Bleaching powder men received far higher wages than the other process men.

The changes in personnel brought about by the introduction of such new sections as the Chance-Claus sulphur recovery plant can be judged from the twenty-sixth report of the Alkali Inspector (1890):

> In one works there are four sets of seven carbonators, twenty-eight in all, connected by a series of pipes provided with cocks for directing the passage of the gases in or out of several vessels. Arranged in a single building there may be counted no less than two hundred and four cocks. The correct position of each of these is essential to the continuance of the operation. Yet so great has been the advance of education generally . . . that it has been possible to select from the ranks of the ordinary workmen those who can be trusted to control this complicated apparatus.

The alkali works, in fact, had become mechanized, and the workers had to learn to control pieces of machinery. The first part of the process to manifest this change was the ball-house where black-ash was made. In 1848 W. W. Pattinson of Felling had devized a furnace in which the melt was mixed and the balls taken out by machinery, but the combination of heat and chemical attack was too much for it: moreover, the balls were difficult to lixiviate in water. In 1853 the first revolving furnace was introduced by G. Elliott and W. Russell, but uneven heating had an adverse effect upon the quality of the balls. Two years later J. Williamson and J. C. Stevenson of the Jarrow Chemical Works patented a change in the mixing process which made the revolving furnace a workable project: they first added the limestone and coal, heating until some quicklime was formed, and lastly put in the ground saltcake.[12] In this way the balls contained sufficient lime for the reaction with water to proceed vigorously, so accelerating the process of dissolution. They also improved the construction of the furnace. Essentially the revolver consisted of a cylindrical barrel, about nine feet in diameter and sixteen feet in length, and fitted with a toothed girdle so that it could be rotated by means of a cog-wheel (Pl. 6). In the middle was a man-hole through which the ingredients could be charged (when the man-hole was upwards) and through which the balls were discharged when the man-hole was downwards. Flame and hot gases from a furnace passed axially along the cylinder. Such a furnace would work fifteen to eighteen tons of saltcake in twenty-four hours, but James Mactear, the brilliant manager for Charles Tennant at St Rollox, built barrel-shaped revolvers capable of working fifty tons of saltcake.

Not only was the engineer invading the Leblanc factory but the scientific chemist was beginning to find a place. The newer techniques called for stricter chemical control, and this in turn led to the development of a whole area of analytical chemistry. Perhaps the two greatest men in this field were Ferdinand Hurter (a former pupil of Bunsen) of Widnes, and Georg Lunge, who became Professor of Technical Chemistry at Zurich. Lunge came to Tyneside in the 1860s with the intention of becoming a chemist in one of the twenty-four factories making alkali. Although he possessed a German Ph.D., he found it difficult to obtain a post; one of the leading manufacturers offered him a pound a week, that is two shillings more than a labourer and less than a skilled process man would earn. Eventually he became manager of the TyneAlkali Company at South Shields, where he devoted himself to a detailed study of every facet of the Leblanc process. He wrote over one hundred and fifty papers, mostly about the analytical chemistry of the alkali trade: these showed particularly his ability to use empirical methods under carefully controlled conditions to yield results of the highest scientific value. His volumes on *Sulphuric Acid and Alkali*, though written after his removal to Zurich as Professor of Technical Chemistry, drew largely on his experience in England and provide detailed information about plant and methods; they also provide source material for historians of the chemical industry. Hurter came to this country from Heidelburg in 1867 to work for Gaskell and Deacon at Widnes: as Lunge worked on the analytical chemistry, Hurter interpreted processes, and the two men collaborated in the writing of *The Alkali Makers' Handbook*, in which were set forth every kind of numerical data for the control of the process.

TABLE 4

MAJOR IMPROVEMENTS TO LEBLANC PROCESS

1827   Gay Lussac's tower for concentrating sulphuric acid.

1836   Gossage's tower for absorbing hydrochloric acid gas.

1839   Pyrites burnt to make sulphuric acid.

1853   Revolving furnace for making black ash.

1858   Henderson's process for copper from burnt pyrites.

1859   Glover's tower for denitrifying sulphuric acid.

1863   Alkali Act which brought Gossage tower into general use.

1866   Weldon's process for manganese recovery.

1868   Deacon's process for chlorine from hydrochloric acid.

1883   Chance's process for sulphur recovery.

Leblanc alkali-making was no longer for the little man; and an event occurred in 1890 which eliminated him from the scene for all time. Faced on the one hand by competition from the ammonia soda process (Chapter Five) and on the other by over-production and inefficiency within their own ranks, the Leblanc soda makers formed an association intended to rationalize the trade. First, the large and powerful firms amalgamated into the United Alkali Company with a capital of £4½ million, large enough to bring in all the rest. Among the giants involved in this combination were Tennant's (at Glasgow and Hebburn), Muspratt's, Gaskell and Deacon's, and Hutchinson's (all of Widnes), and Allhusen's at Gateshead. The press raised a campaign against this 'chemical union', and consequently there was little hope of a public issue of shares succeeding: the United Alkali Company was therefore registered as a private company. The objects of the Company, as published in the *Chemical Trades Journal* were all-embracing:

> To carry on business as manufacturers of chemical products and drugs of all kinds and in all branches of such business; as colliery owners and rock-salt proprietors, miners, brine owners, and white salt manufacturers; as dyers, drysalters, and manufacturers of dyes, stains, colours, varnishes, paints, and pigments; as manufacturers of manure, soap, paper-pulp, paper, glass, bricks, pottery, terracotta and sanitary and disinfecting preparations, coke, cement, and artificial stone; as waterproofers, and india-rubber and leather manufacturers; as millwrights, makers of locomotive engines, waggons, and rolling stock, stone and limestone quarry proprietors, lime-burners, owners of mines of all descriptions, and winners and workers of minerals and mineral oils, and the business of preparing mineral substances for sale or for treatment in manufacturing processes; as metallurgists in all branches; to manufacture and supply gas to the property of the company or the neighbourhood, and, in connexion therewith, to carry on the business of a gasworks company; to carry on any business directly or indirectly connected with the generation, accumulation, distribution, supply, or application of electricity.

The company closed down the less efficient factories and allocated areas of production among the others. It brought temporary relief to a doomed industry, but it was, of course, the first stage in the running down of the Leblanc trade: and it happened in 1891, the centenary of Leblanc's patent which had called into being the heavy chemical industry of Great Britain.

CHAPTER FIVE

More Recent Alkali Processes

A system of chemical manufacture as complicated and economically unstable as was the Leblanc process could only survive in the absence of any better way of making soda. Here the history of technology repeats itself, for as in the case of chamber sulphuric acid a better reaction had long been known but had remained undeveloped for lack of the appropriate chemical engineering techniques, so a more direct route to soda had been known at least since 1838. In that year Harrison Grey Dyar and John Hemming patented a process for making soda by mixing brine with solid ammonium carbonate: the byproduct ammonium chloride was heated with chalk to regenerate ammonium carbonate.[1]

In the laboratory this reaction is very easy to carry out. Indeed, Dr Angus Smith, the first Chief Alkali Inspector, remembered seeing his friend Mr Thom make soda in the palm of his hand by means of this reaction. The transition to industrial working, however, presented many difficulties, not least being the loss of ammonia. Dyar and Hemming built a small works at Whitechapel to exploit their patent; since commercial ammonium carbonate is largely sesquicarbonate, * the product they obtained was sodium bicarbonate, which on heating left a very pure soda.

The solution of ammonium chloride was taken down to dryness and the residue heated in a lead chamber with chalk; clearly, only a proportion of the ammonium carbonate was regenerated in this way, and the partners lost so much money in two years that they abandoned the work. During those two years, however, the works was regularly visited by James Young and by one of Muspratt's sons, so that by 1840 Muspratt was ready to operate a large plant at his Newton factory. This operation, too, lasted only two years, by which time Muspratt had lost some £8000.

* Ammonium sesquicarbonate, $(NH_4)_2CO_32(NH_4)HCO_3$ H_2O. Bath crystals are sodium sesquicarbonate.

Meanwhile the ammonia-soda reaction had been worked in Scotland by John Thom at the factory of Turnbull and Ramsay, Camlachie; it was Thom who had performed the palm-of-the-hand experiment for Angus Smith. The method was to mix salt and ammonium carbonate with a little water, put the resulting pasty mass into a bag, and squeeze out the ammonium chloride solution. Gossage tried it in 1854, and in the same year Gaskell and Deacon began operations at Widnes; their plant produced a few tons per week for two years and cost them about £6000. Similar stories of expensive, shortlived operations could be told about French and German enterprises.

In 1863 the matter was taken up by the Belgian chemist Ernest Solvay, who had gained some experience in handling ammoniacal liquors in a gas works. Solvay was not aware of the immense expenditure of time and money which so many earlier workers had endured without reward: he told Lunge that, had he known all this, he would not have had the courage to devote himself to the task which he completed so successfully. Solvay set up a works near Charleroi in partnership with his brother. The mode of working had by this time changed since the early days of Dyar and Hemming. Ammoniacal brine was now treated with carbon dioxide so that ammonium bicarbonate was formed in solution; Solvay's great contribution was one of chemical engineering, and in particular the provision of a suitable vessel for the carbonating stage of the process (Fig. 9).

In 1872 Ludwig Mond was operating a partially successful sulphur recovering process at Widnes, and news of Solvay's success filled him with anxiety for the future of his own project, being as it was on the edge of Leblanc trade. Accordingly, he hurried over to Belgium, was impressed by what he saw, and acquired a licence to work Solvay's process in England.[2] It is impossible to overemphasize the importance of this event. Not only did it revolutionize the alkali industry in Britain, but it also paved the way for a changed structure of the chemical industry itself. And from it sprang two other branches of the chemical trade.

Mond was born in Cassel in 1839, and received his chemical education from Kolbe at Marburg and Bunsen at Heidelberg. After a long industrial apprenticeship in alkali, wood distillation and ammonia production he came to England to work his sulphur process at John Hutchinson's Widnes factory. His partner in the Solvay enter-

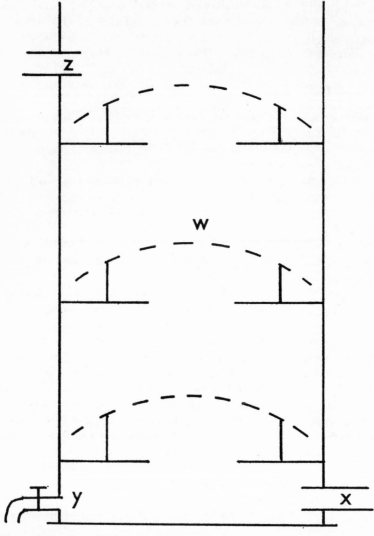

Figure 9 Portion of Solvay Carbonating Tower

w. Perforated dome to disperse carbon dioxide through liquid
x. Inlet for carbon dioxide
y. Stopcock for running off carbonated liquor
z. Inlet for ammoniacal brine

11 Allhusen's Works c. 1910: the last surviving Leblanc works

12 Bleachfields, Monteith's Dye Works, Glasgow, 1842

13 Hawes' Soap Factory, London, 1842

prise was John Tomlinson Brunner, a native of Everton, who was employed in the commercial side of Hutchinson's business. With the aid of several large loans the two men secured a site at Winnington in Cheshire and there commenced in 1873 the manufacture of ammonia-soda. Each partner drew no more than ten pounds a week from the business, and both worked with superhuman vigour. Every kind of technical disaster occurred, and Mond was forced for a time to live on the plant, getting what sleep he could between one crisis and the next.

The heart of the matter was cheap ammonia, and this Mond provided from a producer-gas plant of his own invention. Instead of coke, coal was burnt in an air-steam mixture to yield ammonia, a little hydrogen and carbon monoxide: after scrubbing out the ammonia, the rest could be used as ordinary producer-gas. Having thus shown that coal could be converted in this way, he set up the South Staffordshire Mond Gas Company at Tipton to supply the gas to Wolverhampton and the Black Country; he also used the tar from Mond gas in the preparation of an antiseptic called Monsol.[3]

By the time that Brunner and Mond started production, the Leblanc factories were depending more upon bleaching powder than upon soda for their profits, and so long as the Solvay process could not produce chlorine for bleach, the Leblanc operators could hold out against competition on the soda side. Mond therefore turned his attention to chlorine recovery, at first by treating ammonium chloride with sulphuric acid to produce hydrochloric acid and ammonium sulphate.

In 1886 Mond invented a process for the direct production of chlorine by passing ammonium chloride vapour over nickel oxide, the resulting nickel chloride being thermally decomposed to yield chlorine: the ammonia gas was absorbed in water and returned to the ammonia-soda plant. The process of vaporizing ammonium chloride called for special precautions against corrosion, and as a result of exhaustive laboratory tests, nickel was chosen as the most suitable metal for the valves in this part of the plant. On manufacture, however, the nickel valves appeared to dissolve away in the gas stream and Mond was forced to undertake a fundamental research into the whole matter. It was found that the gases used to sweep out the vaporization plant (thought to consist of nitrogen and carbon dioxide, and obtained from the ammonia-soda towers) were contaminated with carbon monoxide. Furthermore, laboratory studies showed that carbon monoxide combined with nickel at temperatures below 100°C to form a volatile nickel tetracarbonyl which in turn could be thermally decomposed to yield pure nickel. This discovery, in which Mond was assisted by Carl Langer, led to the establishment at Clydach near Swansea of the Mond Nickel Co. in 1931.[4]

Although the growing pains of the Solvay process had brought about the inception of industrial processes for power-gas production and for nickel refining, the troubles of the central organization were by no means over. The original works at Winnington consisted of five Solvay towers (in which ammoniated brine met carbon dioxide), four stills for the recovery of ammonia, a lime kiln, a blowing engine, filters, furnaces for converting bicarbonate to carbonate, a driving engine and a water pump. Water from the River Weaver silted up the boiler tubes and caused an explosion, calcium salts in the brine led to scaling in the Solvay towers, stone in the lime gave rise to troubles in the stills and the blowing engine broke down so frequently that Brunner refused to pay the money owing on it. After a year of operating the plant, Brunner and Mond were producing soda for £16 per ton and selling it at £11. And always there was the problem of ammonia conservation.

At first they bought ammonia in the form of gas liquor at a strength of less than 3 per cent, later setting up a concentration plant in the neighbourhood of their suppliers, Liverpool Gas Company. Gas works and Mond's own power gas plant, provided the necessary ammonia, but the dependence on coal which contains perhaps 1.5 per cent nitrogen was clearly unsatisfactory. Many scientists were concerned about the probable life of British coal deposits: in 1863 Sir William Armstrong, founder of the great engineering and ordnance firm which later became Vickers-Armstrong, used in his Presidential Address to the British Association the following prophetic words: 'The question is, not how long our coal will endure before absolute exhaustion is effected, but how long will those particular seams last which yield coal of a quality and at a price to enable this country to maintain her present supremacy in manufacturing industry.'

Throughout the nineteenth century chemists had at intervals taken up the challenge of converting the inexhaustible supplies of nitrogen in the earth's atmosphere into useful products. Processes based on the formation of cyanides by direct combination of gaseous nitrogen with charcoal or coke had several times been patented, and in 1879 Mond took up one of these in an attempt to increase his ammonia supplies. Barium carbonate was heated with carbon in the presence of nitrogen to form barium cyanide: on treatment with steam, ammonia was produced and barium carbonate regenerated. This process was worked at Winnington but the effect on the plant of the necessarily high temperatures compelled Mond to abandon it after a short trial. As early as 1840 the French chemist Regnault had observed that a mixture of one volume of nitrogen and three volumes of hydrogen when sparked in a eudiometer over sulphuric acid gradually disappeared with the formation of ammonium sulphate: as always, however, there was a world of difference between a laboratory experiment and a successful industrial process. The technical solution of the problem of converting atmospheric nitrogen into ammonia was solved by Fritz Haber.

Haber's training had been essentially that of an organic chemist: born in 1868 in Breslau, he studied under Hofmann in Berlin, Bunsen in Heidelberg, and Liebermann in Charlottenburg before going on to research with Knorr at Jena. Later he worked on topics in electrochemistry and on the thermodynamics of gas reactions: as a direct result of these latter studies he was able in 1908 to demonstrate the synthesis of ammonia in the laboratory. Within a few years Haber's method (the union of nitrogen and hydrogen over an iron catalyst) was in operation at the Badische Anilin and Soda Fabrik at Ludwigshafen, and soon after at the great explosive works at Oppau under the direction of Carl Bosch, ammonia was oxidized to make nitric acid for the manufacture of explosives.

As the first world war drew to an end, the Government prepared for the synthesis of ammonia a site at Billingham-on-Tees. The Armistice was declared before the plant could be got ready, and in 1919 Brunner and Mond took over the site. After visits to Germany, and the gleaning of what slight chemical and technical details they could discover, the Brunner-Mond chemists and engineers settled down to the business of making a pilot plant work. The Billingham site was developed under a Brunner-Mond subsidiary, Synthetic Ammonia and Nitrates Ltd.[5] To this day, old inhabitants of Billingham refer to ICI as 'The Synthetic'. The major portion of the Billingham ammonia was used to make ammonium sulphate for fertilizer, by reaction with anhydrite from the virtually inexhaustible deposits beneath the works. By the end of 1929 the Billingham plant had a productive capacity of nearly a million tons of nitrogenous fertilizer, but by that time the whole structure of the industry had undergone a fundamental change.

The series of scientific crises which the war conditions had precipitated taught lessons which neither Government nor industry could afford to ignore. Larger units with resources to carry out necessary research, economic coordination between the various branches of the chemical trade, the avoidance of overlapping and waste, were seen to be essential features of any postwar structure. Accordingly the four greatest chemical concerns then existing in Britain came together in 1926 to form Imperial Chemical Industries Ltd: these four were the United Alkali Co. (Chapter Four), the British Dyestuffs Corporation (Chapter Nine), Brunner, Mond and Co. Ltd, and Nobel Industries.[6]

Meanwhile Brunner and Mond had acquired majority holdings in various other firms which together represented a wide range of chemical enterprise. Among these was the Castner-Kellner Alkali Company which forms a bridge between the older and new methods for alkali production. The company was formed in 1895 and commenced production at Weston Point, Cheshire, in 1897.[7] Hamilton Young Castner, a native of Brooklyn, New York, was attracted by the possibility of making aluminium cheaply by reducing aluminium chloride with metallic sodium. The chief obstacle to the success of this process was the high price of sodium. In 1854 Deville had made sodium for the aluminium process by the reduction of caustic soda with carbon, chemically sound but manipulatively difficult owing to the tendency of the carbon to float on top of the molten caustic soda. Castner overcame that difficulty in 1879 by substituting for pure carbon a material weighted with iron: he prepared his material by fusing together iron filings and pitch, cooling, and converting to an iron-bearing coke by heating in a large crucible. The metallic coke, when ground to a powder, was sufficiently dense to sink in molten caustic soda. To make sodium, the iron coke was mixed with caustic soda and fused in a crucible at a low temperature. The fused material in its crucible was then transferred to a furnace at 1000°C, the reduction completed and the sodium distilled off (Pl. 25).

In 1886 Castner came to England and made contact with the

Webster Crown Metal Company of Solihull, Birmingham. This firm was buying sodium at 14 shillings a pound and making aluminium at 60 shillings: at this stage Castner offered them sodium at less than one shilling a pound. Two years later a factory was built at Oldbury, Birmingham with a capacity of 100,000 pounds of aluminium per year. The process was also worked at Wallsend-on-Tyne by the Alliance Aluminium Company in 1889. At Oldbury, bauxite (Al_2O_3) was mixed with salt and charcoal and heated in a stream of chlorine to form a mixed chloride of aluminium and sodium: this double chloride was then packed with thinly sliced sodium into a revolving drum for mixing, and the charge introduced into a furnace. At Wallsend a more spectacular process was used in which a lump of sodium was forced into a mixture of fused cryolite and salt. To carry out this dangerous operation, two men stood on boxes and manipulated a hand dipper by means of a rod held between them (Pl. 22).

It seemed at first as if the rewards for working such a process would be high indeed, for, as Curt Netto had pointed out, 'every common brick contains about two to three pounds of aluminium'.[8] But just when this reward seemed within Castner's grasp, C. M. Hall in America and P. Heroult in France brought out a method for manufacturing aluminium by the electrolysis of alumina dissolved in molten cryolite: by 1889 the electrolytic process was being worked commercially and was producing aluminium more cheaply than Castner could. Having a means of making cheap sodium, the only course open to Castner was to develop the manufacture of those sodium compounds best made from the metal. Accordingly he began to make sodium peroxide by burning the metal in a stream of air: whilst discovering the best conditions for this apparently straight-forward reaction, an explosion occurred which caused the collapse of the entire building.

The next venture into sodium salts was the production of sodium cyanide for the MacArthur-Forrest process of gold extraction as worked by the Cassel Cyanide Co. At first Castner used a trouble-some method in which ammonia was passed over molten sodium to form sodamide, and the fused sodamide was poured onto red-hot charcoal. In 1894, however, he patented a single-stage route, in which sodium, carbon and ammonia reacted to form cyanide directly. It is ironical that these sodium compounds were made under the name of the Aluminium Company. Nevertheless the demand for sodium cyanide was so great* that the old process for making sodium metal could scarcely provide enough, and Castner was compelled to look for a better way.

In 1890 he worked out an industrial form of the reaction by which sodium had first been isolated in 1806 by Humphry Davy. Caustic soda was melted in an iron pot, about two feet deep and one foot in diameter, and electrolysis was carried out between a cylindrical nickel anode and an iron cathode which passed up through the base of the pot. The whole plant consisted of fifty cells. Molten sodium rose from the cathode and was ladled out at intervals, more caustic being added to make up the loss. From time to time the cells had to be cleaned out owing to the deposition of impurities arising from the caustic soda. The next step, clearly, was to prepare a pure caustic soda from which sodium might be made in a continuous process, a quest which led to the introduction of the mercury cell for the electrolysis of brine.

The Castner rocking cell was a rectangular box of slate divided into three compartments by means of slate partitions reaching almost to the bottom (Fig. 10). Enough mercury was added to seal off each of the compartments: the middle division was filled with water, the two end ones with brine. Each end compartment contained a carbon anode, the mercury being the cathode. In use, the cell was gently rocked by a cam mechanism so that the mercury circulated, carrying sodium amalgam from the brine compartments to the water in the middle where caustic soda was formed. Castner offered his cell both to the United Alkali Co. and to Brunner and Mond, but as neither was enthusiastic the cell was first developed on a large scale on the Continent and in America: Castner cells continued to be worked in Britain until 1952. It is noteworthy that, among the technical difficulties that he had to overcome, Castner discovered a way to make graphite more durable for use as electrodes. In 1902 the Castner-Kellner Co. began to use a long narrow cell without rocking mechanism, along which the mercury flowed in a continuous stream.

Other workers had found their own solutions to the vexing problem of keeping the caustic soda away from the chlorine which is also formed in the electrolysis of brine. A cell in which a porous dia-

* Gold was discovered in the Transvaal in 1885, and in the Klondyke in 1897.

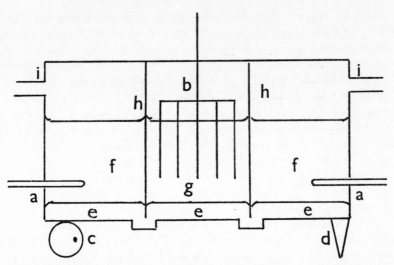

Figure 10 Rocking Cell for Electrolysis of Brine, 1897

a. Anodes
b. Cathode
c. Eccentric
d. Pivot
e. Mercury
f. Brine
g. Caustic soda solution
h. Partitions with free space beneath
i. Chlorine outlets

phragm separated the two products was developed in 1890 by James Hargreaves and Thomas Bird of Widnes. The Hargreaves-Bird cell went out of use about 1927. The popular Hooker cell, which produces a large proportion of the total output of chlorine and caustic at the present time, had its origins in an invention by C. P. Townsend in 1903 and was continuously developed over the next fifty years: Hooker cells were introduced into Britain in 1951. Carl Kellner, the other half of Castner-Kellner, came into the organization on account of his owning a prior patent for a mercury cell in Germany. Rather than embark on a lengthy legal argument in order to gain permission to patent his own cell in Germany, Castner entered into an agreement with Kellner for a mutual exchange of patents; Kellner made very little contribution to the Castner-Kellner Company.

All this concern with the electrolysis of brine arose originally out of the need to supply pure caustic for the electrolytic production of sodium. Attempts to make sodium directly by the electrolysis of fused salt had failed on account of the high fusion temperature required, and consequently the elaborate two-stage procedure seemed the only solution. In 1924 J. C. Downs developed a cell in which a mixture of salt and calcium chloride was electrolysed, the mixture fusing at a lower temperature than the pure salt, and so provided a direct route to sodium metal. The Downs cell came into Britain at the Murgatroyd Salt and Chemical Co. in Cheshire in 1937.[9]

The caustic cells yielded chlorine as a second product, and this was converted to bleaching powder. It will be remembered that the Leblanc soda makers had felt themselves safe from the competition of ammonia soda on account of the inability of that process to provide chlorine for bleach. The growth of the electrolytic alkali trade threatened the United Alkali Co. on two fronts, pure caustic soda and cheap chlorine obtained without the costly capital investment in 'Weldons'. Moreover, the electrolysis of brine produced hydrogen (from the reaction of sodium with water) and so provided a link with the synthetic ammonia industry.

Electrolytic hydrogen was in fact the first source used in the experimental plants to work the Haber process. While coal was still cheap, hydrogen for ammonia synthesis was obtained from the water gas reaction followed by a 'shift reaction' over an iron oxide catalyst.

$$C + H_2O \rightarrow CO + H_2$$
$$CO + H_2O \rightarrow CO_2 + H_2$$

and this source supplied the great synthetic plants at Billingham for several decades. Now that the price of coal makes this process too expensive, hydrogen is produced for ammonia synthesis by steam reforming of naphtha.

CHAPTER SIX

Soap and Bleach

We bring together these two major commodities because of their peculiar position in relation to the chemical industry. Both are cognates of the textile trade and both had profound effects on alkali manufacture. It was this demand for soda to make hard soap (rather than potash which makes soft soap) which provided the main outlet for Leblanc alkali; and later it was the byproduct from soap-making which gave the explosives trade one of its raw materials. Similarly bleaching powder was initially made on a large scale in order to utilize a troublesome waste material of the Leblanc factories, until it came eventually to be the mainstay upon which those factories depended for their economic survival.

The bleaching action of chlorine was demonstrated in 1785 by the French chemist Claude Louis Berthollet, who had recently been appointed inspector of dyeworks and director of the Gobelin establishment. Two years later he showed his experiments to James Watt, who quickly put the process into operation in Glasgow. Other early users of chlorine for bleaching were Gordon Barron and Company in Aberdeen (spinners, weavers and cotton printers), and Thomas Henry in Manchester. Both fabrics and operatives suffered from the effects of chlorine (the French workers had been recommended to chew liquorice to alleviate the symptoms!), and on this ground the practice of dissolving the chlorine in alkaline solutions was quickly adopted. The two solutions were known as 'Eau de Javelle' (in potash) from the name of a bleaching district near Paris, and 'Eau de Labaraque' (in soda) from the name of the Paris apothecary who invented it. In these solutions, only half the chlorine is available for bleaching, the other half being present as inert sodium or potassium chloride

$$Cl_2 + 2NaOH \rightarrow NaOCl + NaCl + H_2O$$

In 1798 Charles Tennant (Chapter Two) introduced the use of

milk of lime, instead of soda or potash, as a vehicle for chlorine. Lime was cheaper than alkali, but all the other disadvantages of bleaching solutions prevailed (including the cost and inconvenience of transport). Nevertheless, Tennant was able, in partnership with Charles Macintosh, to sell 'lime bleaching liquor' from his bleach works at Darnley. A year later he patented the use of dry lime to absorb chlorine, a process worked out by his associate Macintosh. Before 1799 was out Tennant erected works at the St Rollox district of Glasgow for the manufacture of his 'bleaching powder'.[1]

The original preparation involved the decomposition of a mixture of salt, manganese dioxide and sulphuric acid in a leaden still heated by a water bath, the chlorine being absorbed in sifted slaked lime in a leaden receiver. The lead stills were later replaced by vessels of stone, heated externally by means of steam. After the first few years, the question of finding a suitable use for the residual sodium sulphate (salt cake) in the chlorine stills assumed increasing importance: some of it was calcined with sawdust in a rudimentary Leblanc process and used for soap-making, whilst some was sold as Glauber's salt. In 1816 a good deal of correspondence was exchanged with French manufacturers, and on this basis limited Leblanc soda making was pursued, only soaper's salt and residues from the chlorine stills being converted. After 1823 Leblanc soda became, with soap and bleach, one of the major products of the factory.

During the year 1799–1800 fifty-two tons of bleaching powder emerged from the chambers at a price of £140 per ton. Five years later the make had risen to 147 tons and the price had fallen to £112 per ton. Table 5 shows the subsequent pattern of production and price. The effect on the expense of bleaching cloth was immediate and dramatic: in the year 1800 the cost had dropped from seven shillings to sixpence for a piece of twenty-five yards length. Charles Macintosh computed the annual saving to the linen trade of Ireland alone as £166,800.

The St Rollox works grew until it covered fifty acres of ground in 1875: at that time it employed two thousand men, and the output of bleach was about ten thousand tons per year. In 1865 the firm erected works at Hebburn-on-Tyne, with one of the few sites on Tyneside to possess a deep-water quay at that date. Close links were maintained with the Jarrow works of Williamson and Stevenson, the two companies sharing in the development of the South Durham

TABLE 5

PRODUCTION AND PRICE OF
BLEACHING POWDER AT ST ROLLOX WORKS

YEAR	TONS	PRICE PER TON		
1799–1800	52	£140	0	0
1801	96	130	0	0
1802	72	112	0	0
1803	NO FIGURES			
1804	131	112	0	0
1805	147	112	0	0
1810	239	93	0	0
1815	377	81	0	0
1820	383	60	0	0
1825	910	27	0	0
1830	1447	25	0	0
1835	2122	22	0	0
1840	2383	26	0	0
1845	3861	16	0	0
1850	5719	14	0	0
1855	6260	11	0	0
1860	7459	11	0	0
1865	8431	10	10	0
1870	9251	8	10	0

(James Mactear, *Chemical News*, 35, 1877, 23.)

(Haverton Hill) salt deposits, and also in the founding of a local factory to work the copper in the pyrites. Tennant's were fortunate in acquiring the services of John Theodore Merz in their Hebburn copper works (operating under the Tharsis title). This chemist, author, philosopher, electrician, businessman and financier found a useful way to dispose of the waste iron residues, previously dumped outside the factory, and in their way as much of an embarrassment as alkali waste. The red powder was too fine to be handled in any iron or steel plant, but Merz succeeded in briquetting it into lumps that could be treated in conventional steel works: he also persuaded Siemens to take the briquetted ore from Tharsis.[2]

The Tennant works at Hebburn shared with Allhusen's works at Gateshead the distinction of being the last of the Leblanc soda works to operate on Tyneside. After the takeover by the United Alkali Company in 1890 the Hebburn and Gateshead factories were run in conjunction with each other, the manufacture of soda crystals being concentrated at the Tennant works and that of caustic soda at Gateshead. The Hebburn factory was allowed to run down after the first world war: also taken over by the United Alkali Company, the parent works at St Rollox was more fortunate for it survived until 1965. In 1926 when it became part of Imperial Chemical Industries it employed only a hundred men.

The production of bleaching powder was a big undertaking for an alkali manufacturer. He would need lime, chlorine recovered by Deacon or Weldon methods, chambers for absorption, and packing sheds, together with men to operate them. We must now look at each of these in turn. Lime intended for bleaching powder had to be of special purity. The Lancashire makers could command pure and well-burnt lime from Buxton, but most manufacturers found it necessary to burn their own limestone in lime kilns. The Tyne alkali makers also enjoyed an advantage in that pure French limestone (known as 'cliff') was brought back from the Seine by colliers returning in ballast. Cliff was a very pure calcium carbonate which yielded a lighter powder than the Buxton lime: weight for weight, therefore, Tyne bleaching powder required 20 per cent more casks than the Lancashire article. An average kiln held about ten tons of limestone and required five tons of coal for burning, which lasted four or five days and nights. The quicklime was slaked by spreading on a brick floor to a depth of about a foot and sprinkling with a rose attached to a hose: it was necessary for the men engaged in this operation to protect their faces.

The sifted lime was taken to the bleach chambers for spreading. The earliest chambers, after Tennant's lead boxes, were made of wood coated on the inside with tar. For a time these were superseded by chambers of flagstone or slate, but both eventually gave way to brick erections. When, however, bleaching powder came to be the main product of the soda works, brick chambers were found to be too small and there was a return to lead or iron plates: accumulated experience in the vitriol sections had shown how to construct chambers of at least 100 feet in length, and this scale became normal for a bleach chamber.

The chamber was usually mounted on pillars so that bleaching powder could be raked out through trap-doors into barrels placed beneath them. Chambers were usually worked in sets of three, the gas passing from one to another. One of the three was always cut off, either for dressing the lime or for raking out the bleach: in the other two, the strongest chlorine was fed to the most saturated lime.

When bleaching powder was first manufactured, the recovery of waste hydrochloric acid from soda works was not yet practised and chlorine was made from salt, manganese ore and sulphuric acid. The theoretical quantity of chlorine predicted by the equations

$$2 \, NaCl + MnO_2 + 2H_2SO_4 \rightarrow Na_2SO_4 + MnSO_4 + Cl_2$$

was never achieved. The most common apparatus for this purpose was a lead vessel fitted with devices for filling and stirring and protected over its lower part by means of a cast iron shield. All the openings were provided with water-seals because other kinds of joint would corrode too quickly. The residue was often simply discarded, Tennant's use of it in soap making being exceptional.

When waste hydrochloric acid came to be used as a source of chlorine, the bleaching powder maker had to decide whether to use Deacon or Weldon chlorine: his decision influenced the construction of the bleach chambers. For Deacon chlorine, diluted as it was with nitrogen, chambers were furnished with shelves on which the lime could be spread thinly. Where the concentrated Weldon chlorine was used, the lime was spread on the floor of the chamber to a depth of about three inches. Inspection windows of glass enabled the colour of the chamber atmosphere to be observed, chlorine being admitted until a permanent green was established. The Weldon chlorine was generated in octagonal stills made of flagstone impregated with tar: instead of cement for foundations and for jointing, a mixture of tar and sand was used. Such a still of nine feet end diameter cost in 1873 £2,877, including a steam boiler for heating. At the same date the cost of a Deacon plant was about £8,000. This is perhaps sufficient reason why by far the major proportion of English bleach was made by the Weldon process, enough to justify the pronouncement by Dumas that 'by Mr Weldon's invention, every sheet of paper and every yard of calico has been cheapened throughout the world'.

In order to pack the bleaching powder, bleach packers (p.48) had to enter the chamber, shovel the powder into a heap for complete mixing, and rake it out through the trap doors into casks. No matter how carefully the chamber had been cleared of chlorine, the agitation of the powder would liberate fresh quantities of the gas. The recognised English protection (Pl. 16) was a 'muzzle' of twenty or thirty layers of damp flannel tied over the mouth and leaving the nose free: the wearer had to breathe in through his mouth and out through the nose. Continental practice included the use of helmets supplied with air from a bellows outside the chamber but Lunge remarks (and many a frustrated safety officer will agree with him): 'But it would appear that English workmen cannot be got to employ such measures.' The work of a bleach packer consisted of periods of unpleasant labour, of short duration, combined with routine inspection of the chamber and its external fittings: always, however, the risk of gassing was real. The usual remedy for a dose of chlorine was whisky or brandy, for which purpose a bottle was always kept in the office. The United Alkali Company was largely instrumental in introducing into this country the mechanical bleach chamber, invented by Hasenclever in 1888, the only satisfactory solution to the problem of bleach packing.

Soap
We need not concern ourselves here with the various attempts to assign dates to the discovery of soap or to its first manufacture in Britain. So long as soap was required only for domestic purposes, the homemade product would satisfy the market. The impetus to make soap on a commercial scale came from the textile industry, though soap-making has several other industrial connections. While tallow was the most important fatty ingredient, there were associations with the trade in candles for only hard tallow was of use to the candle-makers, who could be led in this way to utilize their waste soft tallow in soap manufacture. Such a connection with candles explains the growth of a soap trade in coalmining areas where there were no textile outlets. In Northumberland, for example, there was much sheep farming to furnish the tallow as well as numerous collieries to consume the candles, and perhaps this is why we read in the *Newcastle Courant* for 1712 an advertisement for 'the English Barrel

TABLE 6

PRIMITIVE CHEMICAL INDUSTRY OF ANIMAL PRODUCTS

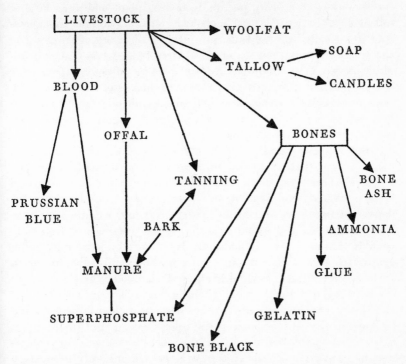

Soap . . . which is sold by retail at threepence a pound, and if not as good as any Crown soap let them return it and they shall have their money again'.

The traditional way to make soap was to boil together oily or fatty material with alkali and whatever else the individual maker might imagine would improve his results; such additives might include sand, rosin, borax, fullers' earth or flour. The chemistry of the soap-making process was investigated by Michel Eugene Chevreul from 1811 onwards. This extraordinary man (who lived to be 103) patiently clarified the nature of oils and fats, showing them to be compounds of glycerol with fatty acids such as stearic, oleic, and palmitic. He further demonstrated that the action of alkali on oils and fats was to liberate this glycerol, at the same time forming sodium or potassium salts of the acids. Once these facts were known, the confusion about useful additives was easily resolved.[3]

The manufacture of soap, then, required oils and fats, alkali (preferably caustic soda), perfumes, boiling plant and some kind of selling organisation. Vegetable oils were at first obtained by expressing from seeds, and the title 'oil mill' in a directory often points to the adjacent presence of soap works. After 1843, when carbon disulphide began to be manufactured by passing sulphur vapour over heated charcoal, the extraction of oils by means of this solvent became widespread. Fish oils, and particularly train oil from whale blubber, were frequently used, and indigenous tallow supplies were supplemented by imports from Russia.

Until the advent of the electrolytic processes, caustic soda was made by causticizing sodium carbonate or by evaporation of the mother-liquors from soda crystallization. After the black-ash had been lixiviated, the alkali waste separated and the solution boiled down until soda crystals formed, there was left a liquid known as 'red liquor'; the red colour was due to the presence of iron (and possibly to thiocyanates in addition). In 1845 red liquors were being evaporated at St Rollox, organic impurities being dispelled by fusion with sodium nitrate. Several leading alkali makers took a hand in improving the process, among then Gossage, Deacon, Gamble and Lee. From 1857 to 1859 the practice became standard of converting carbonate to caustic by adding the calculated amount of lime. Indeed, the mode of manufacture of carbonate was altered by increasing the proportion of lime in the revolvers when it was intended to work up the red liquors for caustic. By 1867 Tyneside was making nearly 4,000 tons of caustic and Lancashire more than 11,000.

Soap was (and often still is) made in pans or kettles; a pan was a shallow vessel presenting a large surface whereas a kettle is deeper. The mixture of fat, alkali and water was brought to the boil, and stirred to form an emulsion in which saponification* could take place. It was known at the end of the seventeenth century that the addition of salt would cause the soap to come out of solution and rise to the top in the form of a curd, but the regular adoption of the practice seems to have come about a century later. The remaining liquid contained salt and glycerol together with any unchanged alkali. At

* Saponification: decomposition of oils, fats or waxes by boiling with alkali, soap being formed.

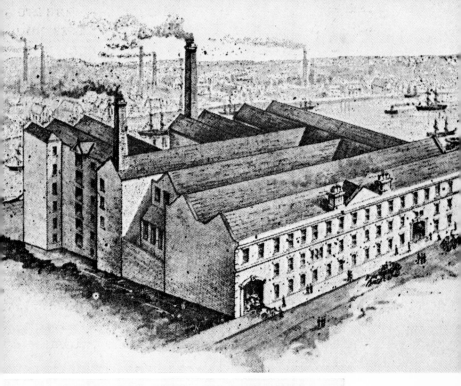

THE BRITISH LION DESTROYS THE GREEDY SOAP TRUST

14 *Above:*
Thomas Hedley
Soap Works, now
worked by
Proctor &
Gamble. 15 *left:*
Public feeling
against the Soap
Trust

16 Bleach packers wearing flannel masks, *c.* 1891

first the glycerol was of no value, and even after 1846, when the Italian chemist Sobrero discovered nitroglycerine, there seemed little possibility of a commercial outlet for it. In 1863 the Nobel family began to make nitroglycerine as an explosive, but a series of disastrous accidents dashed hopes of any early development. Later Nobel was able to show that nitroglycerine could be absorbed on kieselguhr to produce a relatively stable explosive: this he called dynamite, and the first British factory for its production was opened at Ardeer in Ayrshire in 1871. This, and the further discovery of blasting gelatine in 1875, induced the soap makers to recover their glycerol. In the early days of alkali manufacture the aqueous layer underlying the soap was a useful source of salt, bought under the name of 'soapers salts'.[4]

It should be mentioned that whereas alkali makers held banquets in chimney bases, soap boilers held them in kettles. A Newcastle paper for July 1828 records 'On Wednesday evening Mr James Wright, the manager of the soap manufactory in the Close, Newcastle, and thirty-two of the workmen, sat down to an entertainment in the interior of a newly erected soap copper. The health of their employers and other appropriate toasts were drunk with much applause, and the company did not separate till a late hour.'

The first of the large soap firms to become nationally known was that of Andrew Pears who in 1789 began to make a transparent soap in Oxford Street, London. This soap won a prize at the Great Exhibition of 1851. The firm is famous for its pioneering of high-powered advertising, first introduced by its brilliant and courageous director Thomas Barratt. Taking as his motto the fine phrase 'Any fool can make soap but it takes a wise man to sell it', he made use of puzzle posters and optical illusions which caused passers-by to block the pavements. His most striking success was the purchase of Sir John Millais's picture 'Bubbles' which he turned into an advertising poster, but his most daring feat was probably to get the firm's name on the back of the penny lilac stamps of 1881 and the halfpenny vermillion of 1887. It was said that there was a sudden diminution in the number of nude paintings in the Royal Academy for fear lest the picture might be found on a gable end with a soap caption under it.[5]

Of more scientific interest is the firm of Joseph Crosfield founded at Warrington in 1815. Crosfield had served his apprenticeship with

Antony Clapham, the Quaker soap and alkali manufacturer in Newcastle; it is of interest to note that Hugh Lee Pattinson also received his training in Clapham's soapery. The Warrington site was ideally placed for alkali from Merseyside and salt from Cheshire, with an immediate outlet to the Lancashire textile industry. The firm soon made use of Gossage's discovery that sodium silicate (waterglass) could modify the detergent properties of soap; they developed the silicate interest further to make Carbosil (a water softener) and Persil (a bleach depending on oxygen liberated from perborate).[6]

In 1897 Sabatier and Senderens showed that many unsaturated organic compounds (that is, containing double bonds between carbon atoms) could be saturated by treatment with hydrogen under pressure in the presence of a nickel catalyst. Crosfields took up this process, using it to 'harden' unsatisfactory oils into useful fats. Crosfield's firm was bought by Brunner and Mond in 1911, and after the acquisition of the Erasmic Shaving Soap Company it passed into the control of Lever Bros Ltd.[7]

The work of William Gossage in connection with trapping effluent hydrochloric acid gas has already been mentioned (Chapter Four); we must now look at his soap-making activities. Born in Lincolnshire in 1799, Gossage was first apprenticed to a druggist at Chesterfield. When his time was served he set up in business for himself at Leamington, also preparing and selling the salts from that Spa. Salts in general led him to salt in particular, and he moved to Stoke Prior in Worcestershire to commence the manufacture of salt and alkali. In 1850 he started at Woodend, near Widnes, the enterprise that was to become famous for copper, chemicals and, especially, soap. Gossage patented in 1854 his process for 'improvements of certain kinds of soap effected by introducing into soap soluble glass'. The sodium silicate or soluble glass was made by melting together soda ash and clean sand in a reverberatory furnace; when the viscous aqueous solution of sodium silicate was mixed with soap made from tallow or oil there resulted a compound soap with improved detergent powers.[8]

Three years later Gossage patented a modified silicate soap made by boiling together ordinary soap, silicate and colouring matter till such a consistency was reached that the soap would crystallize out with a mottled appearance. This 'blue mottled' soap quickly gained fame, and became the most popular product from the Widnes factory.

Gossage's business, too, was bought by Brunner and Mond in 1911, and, like Crosfield's and Pears, it was subsequently acquired by Lever Brothers.

The possession of a source of raw material is a strong inducement to enter the trade that consumes it. As dealers in hides, skins and tallow Joseph Watson and his sons were in touch with candle and soap makers, and in 1848 they commenced the manufacture of both these articles in Leeds. During the 1880s they produced an antiseptic toilet soap 'Nubolic', which was widely advertised as a prophylactic in epidemics of infectious diseases. They also brought out a scouring powder under the name of 'Watson's Matchless Cleanser'. Perhaps Watson's most valuable gift to the British soap trade was their employment, in 1887, of Dr Julius Lewkowitsch as their chief chemist. Lewkowitsch, who had been assistant to Victor Meyer at Heidelberg and manager of a tar distillery at Brunswick, quickly established himself as an expert on the physical and chemical properties of oils and fats. He published the monumental *Chemical Technology and Analysis of Oils, Fats and Waxes* which became the standard work of reference on this subject. He was also the first to isolate chemically pure glycerol from the lyes.

In 1906 the firms of Watson and Gossage combined and in the same year they joined Lever's Soap Trust, designed to cut the costs of competitive advertising and pilloried by the *Daily Mail* as an attack on the housewife[9] (Pl. 15). William Hesketh Lever differed from Crosfield, Watson and Gossage in that he sold soap before he made it. In connection with a grocery business inherited from his father, he marketed a palm oil soap, not in bars from which the retailer could cut off the required amount, but wrapped and in tablet form: this was Sunlight Soap and was made for Lever by Watson and Gossage. In 1885 Lever bought up a small soapery at Warrington and over the next two years developed it from a twenty tons per week business into one of 450 tons per week. In 1888 he founded the Port Sunlight factory and there began the series of promotions that were to lead to Lifebuoy Soap, Lux and Vim.

In the early years of this century Lever embarked on a programme of acquiring the sources of his raw materials, through palm oil concessions in Africa and whale oil establishments in the Antarctic. There followed transport undertakings and subsidiaries in timber and fishery. At home Lever began to buy out rival soap firms until he possessed Pears, Crosfield, Watson, Gossage, Knight (of Knight's Castille), Gibbs, Erasmic and Icilma. The only serious competition for oily and fatty raw materials was the edible fat trade, and consequently a merger took place in 1929 with the Dutch margarine firm of Jurgens. Thus was born the Unilever organization, with ramifications far beyond the fields of soap and chemicals.[10]

CHAPTER SEVEN

Fertilizers

The manufacture of chemical manure meant strictly the treatment of calcium phosphate with sulphuric acid to produce the more soluble calcium superphosphate, according to the method suggested by Liebig in 1840 and introduced into this country by John Bennett Lawes. Several great chemists had occupied themselves with the connection between chemistry and agriculture, among them the Earl of Dundonald and Humphry Davy. Their experiments resulted in such improved practices as the addition of lime to soil and the extended use of such natural manures as Peruvian guano, but they lacked a coherent theory. In 1837 Liebig outlined some proposals on the subject of mineral requirements of plants at the British Association meeting in Liverpool, and three years later he published his *Chemistry Applied to Agriculture and Physiology*.

J. B. Lawes was born in 1814 and in 1832 he inherited the family estate at Rothamsted in Hertfordshire. Following upon a conventional education for a country gentleman's son he attended a course of lectures in chemistry in London, in this way preparing himself for the task of putting into practice the suggestions of Liebig. In 1841 Lawes set up a works at Deptford which was to be prototype of all superphosphate factories. At first the only available phosphate materials were bones, bone ash, spent animal charcoal, and the poorer grades of guano; these he later supplemented by coprolites (fossilized dung) from Cambridgeshire and Suffolk, first shown to Liebig by the geologist Buckland. In 1857 mineral calcium phosphate was imported from Norway for Lawes's factory. For a time Lawes engaged in the manufacture of citric and tartaric acids at the Atlas Chemical Works, Millwall, which he bought in 1866.[1]

The process of decomposing phosphatic material was patented by Lawes in 1842; he specified 'bones, bone ash, bone dust and other phosphoritic substances, mixing a quantity of sulphuric acid just sufficient to set free such phosphoric acid as will hold in solution the undecomposed phosphate of lime'. The product was a pasty mass and needed to be mixed with dry absorbent matter if it was to be handled conveniently. A patent taken out by Dr James Murray of Dublin in the same year as Lawes describes these absorbent substances as 'bran, sawdust, dust of malt, husks of seeds, brewers' and distillers' grain, ground rags, pulverised rope or linseed cakes, the refuse of flax leaves, bark, dry tan, siliceous sands, peat or other sandy mould, dry dust, earth or clay, fine sifted cinders, ashes and the like'.[2]

The manufacture of superphosphate did not at first call for elaborate plant or machinery. A tank for sulphuric acid, a shed in which to store raw phosphate, another for finished superphosphate and a 'den' for mixing were all that was required. The den might be made of brick coated with tar, or of cast iron embedded in the ground, or it might be improvised as described by Muspratt in 1860.

A circular wall, in size about two feet, is formed by ashes. The space enclosed is about ten feet. The crushed bones are passed through a fine 4 inch sieve. The coarser particles are laid flat in the centre and the finer are placed round the ring, close to the ashes. Water is then added to the bones so as thoroughly to saturate them, after which they are turned over frequently in the course of a few hours. . . . After the bones have been turned the acid is added gradually, the bones being constantly turned and mixed with the spade. In six or eight hours after the addition of the acid, the fine bone dust forming the second circle is thoroughly mixed with the bulk. Next day the ashes forming the outer wall are thrown over the heap which is left undisturbed for a week. The heap is then opened out and thoroughly mixed with the ashes, and finally passed through a sieve.[3]

When a more permanent den was used it was sometimes thought preferable to add the phosphate in small quantities to the bulk of acid already in the den.

When oil of vitriol is poured on to mineral phosphate a vigorous action ensues and corrosive vapours are given off: these can include hydrogen fluoride (from associated calcium fluoride) and silicon tetrafluoride. Further pollution of the atmosphere arose when the pasty mixture was dried prior to packaging. Chemical manure works were therefore brought under the control of the Alkali etc. Works

Act as amended in 1881. The report of the Chief Alkali Inspector makes it clear, however, that acid fumes were only a small part of the price to be paid for having a manure works for a neighbour. 'It not infrequently happens that in some of the works registered for the manufacture of chemical manure, some allied trade is also carried on which is of a nature to give rise to much complaint owing to its noxious or unpleasant character, such as that of horse slaughterer, bone boiler, blood drier, glue maker etc.' This general picture is supported by reports on individual works: 'This is a manure and horse-slaughtering work', 'The chief business here is horse-boiling', and especially, 'At this work the town excreta from Blackburn is manufactured into a manure by a somewhat complicated process, and the smell was by no means insignificant or agreeable'.[4]

The mode of working these allied trades seems to have left a great deal to be desired. Of one such works on Tyneside it was stated:

> The manure is made from carcases, shoddy, leather, slaughter-house refuse, and some mineral phosphate. The method is as follows: a heap of twenty or thirty tons of shoddy is made in the shed, and on this is poured blood and refuse from the slaughter-houses. Any carcases that the owner may buy are, after being skinned, buried in this heap, the heap being allowed to stand and rot for five or six months. This is then shovelled into the mixer with some leather, some crushed bone, and some acid. After mixing this is let into an open den, and a man shovels onto it a certain quantity of ground mineral phosphate. . . . The stench when these mixings are going on is simply intolerable.

In a Lancashire works, pressed cod livers (from making cod liver oil) were put into a heap and left for some months: another is described as making manure from flesh which has been boiled to extract the fat, this residue being mixed with shoddy. A Scottish works was said to be 'An animal manure work. Very offensive. No houses near except those used by workmen', and in its neighbour 'Dissolving done in tubs. . . . Smell from the operations and from the decomposing animal matter very offensive.' It can be seen from this that control by the alkali inspectors was nearly impossible: the only part of the operation to come under their jurisdiction was the mixing of mineral phosphate with sulphuric acid, and all they could offer in respect of the rest was friendly advice.

The scale of manufacturing, too, varied so widely from one works to another that compliance even with the limited letter of the law was difficult to enforce. In 1882 the Chief Inspector stated that it was difficult to treat every works in the same way, and at the same time to be reasonable and just. Some of the establishments were of only a part-time nature and were reported as working only about six months in the year, or only two days in the week: more than once the Inspector found the whole works shut up and no one to answer his questions. The small works especially found it hard to keep up with requirements: 'A very small work indeed. Superphosphate is made in a hole in the ground and no machinery of any kind used for mixing. The owner, a working man, thinks it very hard that he should have to pay a fee of £3 which he says is a very large percentage of his profits.' Probably in the same situation was the man whose works were in the backyard of his house, five miles from the nearest town. The operative feature was the use of oil of vitriol: if this was employed, the works had to be registered, but if the proprietor was content to buy superphosphate and simply mix it with organic matter he need not register. At the other end of the scale there were the large factories such as Lawes's producing 40,000 tons of super-phosphate in the year.

The larger factories were obviously more likely to receive the attentions of the inspectors and for them the condensation of fluorine-bearing fumes was important. In 1881, when these works first came under the Act, there was no accepted method of dealing with the problem. Two works were cited as models in this respect; not—it was carefully stated—that these were perfect, but that any that were worse than these must improve. The model factories were Barnard, Lock and Co., at Plymouth, and John Cran of Inverness. The first procedure to be recommended was one familiar to the older alkali makers, a flue of about four square feet in cross section and at least two hundred feet in length; this was to be followed by a tower in which solution of the gases in water might take place. One of the most successful towers was designed by John Morrison of Newcastle, and since his works was in the heart of a town that was well known for nuisance prosecutions we may assume that the Morrison Tower was efficient. The Alkali Inspector was not disposed to publish re-commended designs, on the ground that he would rather see inven-tion beginning afresh 'so that we may have some novelties in these works instead of the humdrum, either of carelessness or of imitation'.

So long as bones remained the chief raw material, crushing and grinding were easy. Ordinary grinding stones or crushing mills would deal also with coprolites, but the introduction of the very hard phosphatic rocks called for other means of pulverization. The powerful Sturtevant mill was capable of crushing and powdering up to 300 tons of rock every twenty-four hours: the powdered rock was subsequently sieved down to 2000 to 3000 meshes per square inch, to make for easier solution in acid.[5] The handling of phosphate in the form of such fine powders highlighted the inadequacy of manual mixing methods, and consequently the primitive tools, harrows, rakes, shovels and hoes, began to give way to mechanical mixers. These at first took the form of hand-operated paddlewheel devices, but most of the larger works soon adopted a 'revolver' similar to that used in black-ash manufacture. Many of the works made their own sulphuric acid, but others constructed large lead-lined tanks in which to store bought acid. We can now form a picture of a typical fertilizer factory as possessing a vitriol plant or storage tank, crushing machinery, some kind of mixer, sheds for untreated phosphate and for finished superphosphate, and perhaps a yard in which took place the kind of noisome processes that the inspector could not touch.

A chemical manure works might well handle chemicals other than phosphates. Sulphate of ammonia was often made, leading to a liaison between coke ovens and manure works, or an excess of sulphuric acid might lead into general alkali works processes. Typical of that kind of enterprise was the County Durham works of Thomas Richardson founded in 1844. Richardson had been a pupil of Liebig in Giessen and therefore had learned the great organic chemist's views at first hand. For a time the works passed into the hands of a firm of leather manufacturers, perhaps in order to use their tanyard waste in an economical way. In 1877 it was reorganized by J. T. Merz (who had shown Tharsis how to utilize their iron waste, Chapter Four) and acquired the significant title of Blaydon Manure and Alkali Co. Ltd.: this works, finally taken over by Fison's, had the last lead chamber to function on Tyneside (Pl. 19).

Not usually prepared in the superphosphate works, but certainly a chemical manure, was that offshoot of the steel plants known as 'basic slag'. The iron ores on which the great iron and steel trade of Middlesbrough depended suffered certain disabilities due to the presence of phosphorus. The researches of Sidney Gilchrist Thomas and his cousin P. C. Gilchrist succeeded in treating this ore in a Bessemer converter in which the lining consisted of a lime-magnesia brick: the local dolomites provided the raw material. The phosphorus in the iron was converted to calcium phosphate, and the basic slag contained the equivalent of about 25 to 40 per cent. The slag was ground to powder, sieved and sold as 'Thomas Phosphate Powder' or, less frequently, treated in the superphosphate works as a calcium phosphate source.[6]

Apart from the introduction of more refined machinery, and such modifications to the plant as have been dictated by changes in the grades of mineral phosphate available, the manufacture of super-phosphate has undergone few fundamental alterations. It remains one of the major applications of sulphuric acid. The production of nitrogen fertilizers is described in Chapters Five and Eight.

CHAPTER EIGHT

Coal Tar Chemicals

There passed through Britain in the seventeenth century a procession of German technologist-economists who were known as projectors. Their projects had usually to do with exploiting the natural resources of a district for commercial purposes and to them we owe a number of technical innovations. One of these projectors was Johann Joachim Becher who came to England to study the metalliferous mines, especially in Cornwall. In August 1681 he took out a patent for the preparation from coal of a tar which he claimed to be as good as Stockholm tar from wood. He demonstrated his process before the King at Windsor, but his commercial gain can be estimated from the fact that in the following year he died in an almshouse, and his grave has never been found.[1]

Exactly a hundred years after Becher's patent, Archibald Cochrane (the colleague of William Losh in alkali making, Chapter Three) was granted a patent for 'his Invention of a Method of Extracting or making Tar, Pitch, Essential Oils, Volatile Alkali, Mineral Acids, Salts, and Cinders from Pit Coal'. Coke ovens were erected at Culross and with the support of some Newcastle businessmen the British Tar Company was founded.[2] Although the coke could be sold to iron-masters and the ammonia to makers of sal-ammoniac, Cochrane's hopes were really founded on the use of coal tar in treating ships' bottoms, and to that end he published in 1785 a pamphlet entitled *An Account of the Qualities and Uses of Coal Tar*. The Admiralty, however, preferred copper-bottoming, which was good for another indigenous industry but bad for Cochrane who, like Becher, failed to find riches in coal carbonization.

Others had looked to coal for another kind of technical betterment. In 1667 the *Philosophical Transactions of the Royal Society* carried a description by Thomas Shirley of 'a well and earth in Lancashire taking fire by a candle approached to it'. Shirley observed that the place where this uncommon phenomenon occurred was not more than forty yards distant from the mouth of a coal pit and concluded that the flame was fed by 'the erruption of some bituminous or sulphureous fumes'. About twenty years later the Rev. John Clayton wrote a letter to the Royal Society in which he took Shirley's observations a step further: this letter was not published until 1744, fifteen years after the death of its writer. After confirming the existence of the burning ditch near Wigan, 'the flame of which was so fierce that several strangers have boiled eggs over it: the people thereabouts indeed affirm that about thirty years ago it would have boiled a piece of beef', he went on to describe how he had collected coal from the neighbourhood and distilled it in a retort in an open fire. He collected in ox bladders the inflammable gas that issued from the mouth of the retort and used the bladders to divert visitors to his house by pricking with a pin and compressing the bladder near to a candle flame.

There were several tentative experiments in gas lighting during the eighteenth century. George Dixon of Cockfield in County Durham, for example, succeeded in lighting a house in addition to making coke and tar about 1760, but his plant was finally wrecked by an explosion. Archibald Cochrane, in the coal tar project mentioned above, lit up the hall at Culross Abbey in 1787, but these ventures were in the nature of sideshows. So far as Britain is concerned, the pioneer of successful gas lighting was undoubtedly William Murdoch, a strange young man from Ayrshire who on occasions wore a wooden hat of his own making. While employed by Boulton and Watt to erect engines in Cornwall he lit up his house at Redruth in 1792. It is not our purpose to trace the history of gas lighting, but the gas works are important to our story in so far as their byproducts became the raw materials of the chemical industry.

In 1798 Murdoch built a small gas plant at the Soho (Birmingham) works of Boulton and Watt, using a gas holder for storage over water: in 1802 the factory was illuminated by gas to celebrate the Peace of Amiens. Murdoch's assistant, Samuel Clegg, having left Boulton and Watt to join the notorious Winsor in theatrical displays of gas lighting, discovered in 1805 how to purify coal gas by means of lime. Winsor raised enough money to light part of Pall Mall in 1807 and went on to found in 1812 the London and Westminster Chartered Gas Light and Coke Company (later the Gas Light and

Coke Co.), with Clegg as chief engineer.[3] Another chemist who
joined the board of that company in its early days and helped to
publicize the facts about gas lighting was Friedrich Accum. A native
of Hanover, a consulting chemist, analyst and private teacher, a
former assistant at the Royal Institution and lecturer at the Surrey
Institution and a formidable opponent of food adulteration, Accum
wrote in 1815 a *Practical Treatise on Gas Light*. Gas lighting
advanced with great rapidity, and industrial archaeologists who wish
for information about specific towns should scan the Acts of Parliament (Chapter Twelve) for the first quarter of the nineteenth
century.

There was a steady demand for ammonia which gas works produced, but the tar was often looked upon as an inconvenient waste
product. In some places tar was mixed with coke or coal dust and used
for fuel in heating ammonia stills.[4] Charles Macintosh bought all
the ammonia and tar that the Glasgow gas works produced, using
the ammonia in his cudbear* establishment. The tar was separated
into 'spirits' and pitch, at first by Macintosh himself and after 1822
by the firm of Longstaff and Dalton at Leith; the pitch was burnt to
make lampblack, and the spirit was used to dissolve rubber as part of
the process for making waterproof cloth. There is no doubt that much
tar was poured into rivers at this time, and the Gas Light and Coke
Co. were more than once in trouble for polluting the Thames. In
1838 John Bethell used coal tar for preserving railway sleepers, and
in this way the profitable utilization of tar became linked with the
expansion of the railways: much naptha was, of course, used for
burning in flares.

The development of the coal tar trade as a source of raw materials
for the chemical industry in Britain might never have taken place
had it not been for the founding (for quite a different purpose) of the
Royal College of Chemistry. In 1842 Liebig (Chapter Six) had made
a tour of Britain during which considerable attention was focused
on his ideas on chemistry in relation to agriculture. In this way a
favourable climate of opinion was created for the setting up of an
institution devoted exclusively to chemistry and chemical education.
Wealthy landowners put up money, the Prince Consort lent his
prestige, and Sir James Clark, physician to the Queen, together with
Sir Lyon Playfair prepared the blueprint. In the 1830s Liebig's

* Cudbear: a dyestuff extracted from lichens.

laboratory at Giessen had been the Mecca of all ambitious and gifted students of chemistry, and it was to that source that the organizers of the new College turned for a director. After some hesitation, the post was offered to one of Liebig's most brilliant students, A. W. Hofmann, and the Royal College of Chemistry opened in temporary laboratories in George Street, Hanover Square, in October 1845. In the following year a move was made to new buildings in Oxford Street and here Hofmann worked and taught for nearly twenty years.[5]

Hofmann's research interest in Germany had centred round the organic bases present in the liquid naphtha derived from coal tar, and he developed this line of research in London. One of his first pupils, Charles Blachford Mansfield, devised methods for separating pure aromatic hydrocarbons from coal tar naphtha. His starting material was the light oil fraction which he washed with acid, alkali and water before fractioning several times, and his final products included benzene and toluene. Benzene had, of course, been discovered as early as 1825 by Michael Faraday in the course of an examination of illuminating gas made from whale oil, had been recognized in coal tar by Leigh in 1842 and unequivocally identified by Hofmann in 1845. Mansfield used an apparatus commonly employed for rectifying spirits of wine, but the benzene so obtained was always contaminated with toluene: his method of final purification was to freeze the benzene and press the toluene out from the solid, for which purpose he used 'a very simple apparatus—one of Beart's coffee pots'.

By 1849 Mansfield was collaborating with Read Holliday of Huddersfield in an enterprise for supplying benzene for lighting sets on various scales from table lamps up to district plants. He also prepared aniline for Hofmann's further work, presumably by nitrating benzene and reducing the product. Dr Ward of Bournemouth has recently shown that Hofmann regarded the distillation of coal tar hydrocarbons on a large scale as too dangerous for the laboratories of the Royal College of Chemistry, and perhaps for this reason Mansfield set up a small laboratory in St John's Wood.[6] Whatever the background to this, Mansfield was engaged in preparing samples of aromatic hydrocarbons for the Paris Exhibition of 1855 when his still caught fire, and he was so badly burnt that he died a few days later. The hope of exploiting the riches of coal tar, rather than the manufacturing techniques that he employed, was Mansfield's great gift to the chemical industry: some of the possibilities were outlined in his paper to the Chemical Society in 1849.

The solvent properties of benzene were described thus: 'Benzole dissolves many substances with extreme readiness and in large quantities, such as many resins, mastic, camphor, wax, fatty and essential oils, caoutchouc and gutta percha.' One of the immediate uses to which the newly available solvent was put was the removal of grease spots from clothing. Hitherto turpentine had been used for this purpose (most necessary in days when candles and oil lamps were commonly in use) but the smell of turpentine was slow to disappear from the cleaned garment. In addition to laying the foundations of the drycleaning industry, Mansfield made a contribution to the soap trade; his method for making nitrobenzene from benzene gave to the trade a synthetic almond perfume, widely used under the name of 'essence of mirbane'.[7]

Possible medical and surgical uses were not overlooked. Solutions of rubber or gutta percha might be spread on glass so that

a film of the gum is deposited, which may readily be peeled off in the form of a tough membrane of any required degree of tenuity, and possessing all the properties of the original material. The same solutions varnished on the skin form admirable artificial cuticles, which has been found useful in cases of wounds and burns, and might probably be very beneficial in some skin diseases.

Dr Snow, the authority on anaesthetics, tried the effect of benzene vapour on mice and found that 'it causes insensibility, but accompanied with convulsive tremors'.

Mansfield's tar came from gas works and was indeed defined as 'the black oily matter produced in the distillation of the varieties of bituminous coal, at high red heat, for the production of illuminating gas, as it is ordinarily practised in gas-works'. The vast amounts of coke produced for metallurgical purposes were made in ovens that did not allow for the recovery of byproducts. The British coking coals, especially in County Durham, were ideally suited to the old beehive ovens from which came the coke that was pleasantly silver in appearance and worked well in blast furnaces. The capital cost of a beehive oven was low. On the other hand coke made in early byproduct recovery ovens did not look well, lacked the sonorous ring, and cost a good deal more.

The problem, then, was to reconcile the demands of the users of coke and of coal tar. In 1874 Henry Aitken of the Falkirk Iron Works devised a plan for taking off tar and ammonia liquor through the floor of a beehive oven, and this was improved upon by John Jameson of Newcastle upon Tyne in 1883. By that time, however, there had been erected at Crook in Weardale the first battery of byproduct recovery coke ovens in Britain.[8] These were built on the Simon-Carvès pattern, twenty-five in number, and each measured 23 by 6½ by 19½ feet. The 4½-ton charge of coke was ready in sixty to seventy-two hours. Each ton of coal yielded about six gallons of tar and twenty-eight gallons of liquor. The cost of the coke was significantly higher than that from beehive ovens. Dr Mott records that the manager of this historic battery was paid a good deal less than the engineers: this brings echoes of Goethe's meeting a century earlier with Stauf, the manager of the Dutweiler works near Saarbrücken, who was described as haggard and worn-out, with a boot on one foot and a slipper on the other.

We are not concerned with the development of coke ovens, nor with the breakdown of the prejudices of such notable iron-masters as Sir Isaac Lowthian Bell to the use of byproduct coke. It is sufficient to state that byproduct works increased in number until they were able to provide a sizeable proportion of the coal tar for the organic chemical trade. The quality of the tar changed also, more anthracene occurring in the Simon-Carvès tar than in gas tar, and that at the very time when the dyestuff trade began to demand it.

In the early days of tar distilling, any iron vessel such as an old steam boiler was made to serve, but frequent repairs, difficulty of cleaning and occasional frothing over, all pointed to the necessity for a properly designed still. The shape favoured by the English distillers was an upright cylinder of nearly equal height and diameter, with a dome-shaped top and a bottom made concave like a wine bottle. The material of manufacture was usually wrought iron boiler plate, up to half an inch in thickness; the capacity of the still ranged from six to twenty tons. The still was set over the fire in such a way that the flame impinged on the bottom and the hot gases travelled round the still in an annular flue before going into the chimney. The condensing apparatus consisted essentially of a 'worm' of cast

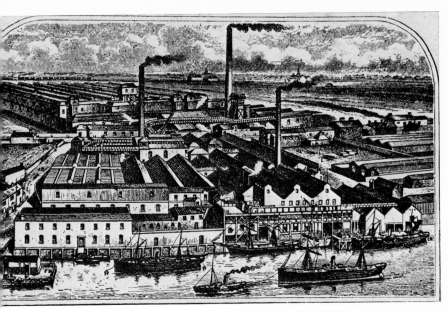

17 Langdale's Manure Works, Newcastle, 1875

18 Tar and ammonia recovery, R. Heath & Sons, Norton, Staffordshire, 1884

19 Advertisement for Blaydon Alkali Works (founded 1843), 1930

iron piping immersed in a bath of water (kept warm with steam for the later stage of distillation): the receivers were changed manually as distillation proceeded.

The fractions collected have been variously named and the temperature boundaries between them have varied similarly. Very roughly, however, the scheme that was generally employed divided the distillation products in this way:

First runnings	up to 110°C
Light oil	110 to 170°C
Middle, or carbolic, oil	170 to 230°C
Heavy, or creosote, oil	230 to 270°C
Anthracene, or green, oil	270 to 330°C
Pitch	Residue above 330°C

Each fraction was subjected to chemical washing (e.g. with sulphuric acid) and to further fractionation. The first runnings contained the benzene, toluene and xylene, together with many minor constituents. The prime chemical value of these lay in the preparation of intermediates for the dyestuff industry: in 1880 it was estimated that German colour works used in one month 1,020 tons of benzene, nine-tenths of which came from the English tar distilleries. Benzene supplies were later augmented by the recovery of it from coal gas. Before Welsbach's discovery that rare earths would glow white-hot in a gas flame and his subsequent invention of the gas mantle in 1886, gas was sold on a basis of illuminating power rather than on calorific value, and there was every incentive to keep the benzene in the gas. Apart from the chemical uses of this fraction of the tar distillate, solvent naphtha was sold to the rubber manufacturers without further separation. The last of the useful products in this group was burning naphtha, used without wick or chimney in flares for open-air illumination at fair grounds, street markets and in factories instead of gas.

The light oil, so called because it is less dense than water, was in demand as a solvent for the more expensive resins such as sandarac, mastic and copal: the superior varnishes for photographers and printers were prepared in this way. The carbolic or middle oil was valued for the phenol and the naphthalene which it contained. Derivatives of phenol were of use in preparing azo-dyes, and the carbolic acid itself was used in making picric acid. Most of the other uses centred round its antiseptic properties: disinfecting sewers, stables and sometimes even human dwellings, treating such cargoes as bones and hides for long sea journeys, and of course in Lister's famous spray for use in operating theatres. Lunge states that phenol was recommended for the preservation of meat 'but does not appear to be practicable owing to the very tenaciously retained smell and taste of carbolic acid'; substantial quantities went to the makers of carbolic soap and antiseptic lozenges. There were several proprietary forms on the market and Professor Crace Calvert of Manchester was not above putting out Calvert's Carbolic Acid Powder (containing 15 per cent phenol in a china clay base).

For a long time the naphthalene was regarded as an unpleasant intruder on account of its crystallizing out in inconvenient places. The two developments in dyestuff chemistry which rescued naphthalene from lowly use as a fuel were, first, the employment of the naphthols in azo-dyes and later the preparation of phthalic anhydride on the route to indigo and the phthaleins. Its use for repelling insects, particularly moths in furs, dates back to 1868. Heavy oil was used mainly for wood preservation or for burning; about the rectification of this fraction for chemical purposes Lunge (1882) said that the expenses of fuel, wear and tear, wages and unavoidable losses seemed to deter most distillers from pursuing it. The timber pickling processes had not changed in outline since Bethell's patent of 1838: the wood was exposed to reduced pressure to draw out moisture and gases, and creosote was then applied under increased pressures.

The burning of heavy oil to make lampblack was carried out in a number of ways. The basic principle was to admit enough air for only partial combustion and to arrange for some separation of the product into the different grain sizes. In a typical plant, a fine stream of oil from a tank was allowed to fall onto a red hot iron plate forming the bottom of a brick chamber: a side door with small holes admitted the necessary air. The smoke passed through four brick chambers in series on its way to the chimney, and when all the oil had been consumed the chambers were opened and the soot scraped out. The last chamber (nearest the chimney) contained the finest black suitable for lithographers, the next a slightly coarser black for printers and the first two the common lampblack for bootblacking and blacklead for stoves. On Tyneside fine lampblack was made by what seems to our eyes to be a most wasteful process of burning turpentine.

Anthracene, or green, oil came into prominence when synthetic processes for alizarine were developed from 1868. The tar distillers had first to be convinced that raw anthracene was worth saving and then to be shown how to do it, and both these objectives were achieved by W. H. Perkin (Chapter Nine). Sir Henry Roscoe in a lecture to the Royal Institution in 1886 pointed out the effects upon the coal tar industry of Perkin's perseverance: a greasy material which in 1869 was burnt in the furnaces or sold as a cheap waggon grease for a few shillings a ton was by 1871 pressed into cakes and sold at a shilling a pound.[9] A French anthracene maker even approached the Paris authorities for permission to dig up the asphalt in the streets so that he might distil it for anthracene. Filter presses were specially designed for this process, and it is worth noting that the chemical utilization of coal tar products brought into being a chemical plant industry of far from negligible proportions.

The residue of pitch in the tar stills was used on the roads, for treating wood and brick used in corrosive chemical storage and manufacture, for making asphalt paper (used since 1870 for papering damp walls), as a substitute for coal in the black-ash process for making soda and, after briquetting, in the manufacture of patent fuel. The stage was now set for the dramatic episodes which so captured the popular imagination, when colours of previously undreamed-of brilliance, drugs and medicinals, perfumes and flavourings, and explosives for peace and war were all to be made from coal tar. Mr Punch had his finger on the pulse of general feeling when he wrote:

> Oil and ointment and wax and wine,
> And the lovely colours called aniline,
> You can make anything from a salve to a star,
> (If you only know how) from black coal-tar.

CHAPTER NINE

Dyestuffs

The synthetic dyestuff industry makes a pleasant change for the industrial historian because, unlike so many other situations dealt with in this book, it is possible to assign to it a definite beginning. The first commercially useful synthetic dyestuff was made by William Henry Perkin in 1856, in a room roughly fitted out as a laboratory, and when he was just eighteen years of age. Perkin was born in 1838 at Shadwell in the East End of London where his father was a builder. At the City of London School he came under the influence of Thomas Hall, a former pupil of Hofmann at the Royal College of Chemistry (Chapter Eight) and one of the first teachers of chemistry in a school. Although the boy was intended by his parents for the profession of an architect, Thomas Hall persuaded them to send him to the Royal College of Chemistry so that he, too, could work with the great Hofmann.[1]

It was the custom of Hofmann, as of all the German chemistry teachers of that time, to put his students to research as soon as they had completed a preliminary course. So in 1854, after a year of instruction, Perkin carried out a research on anthracene (known in those days as 'paranaphthalene') and followed this with a joint investigation with another student on certain naphthalene derivatives. At this stage, and while he was still only seventeen years old, Hofmann made him a personal assistant in his own research laboratory. Hofmann's concerns left him little time for private work, and for this reason the private laboratory was equipped at home; in fact it was probably no less adequate than any laboratory in the College, for fifty years later Perkin stated: 'There were [in the College] no stink closets except the covered part of a large sand bath heated with coke. There were no Bunsen burners but we had short lengths of iron tube covered with wire gauze.' In his home laboratory, although there was no water laid on, he at least had spirit lamps for heating.

The research that Perkin undertook in the Easter vacation of 1856 involved the oxidation of allyl-toluidine with dichromate, from which the product was a reddish-brown precipitate. It has been said that chance favours the prepared mind, and Perkin's discovery owes as much to persistence as to good fortune. For he decided to eliminate some of the complications of his topic by repeating the experiment with aniline (actually contaminated with toluidine), and this time he obtained a black instead of a brown mass. Extraction of this with spirit gave a mauve solution which later proved to have the properties of a light-resistant dye. In June of the same year he sent to Pullar's of Perth samples of silk dyed with his mauve and received comments so satisfactory that he left Hofmann's service and embarked, with financial support from his father, on the manufacture of the first 'coal tar' dye.

It is necessary at this stage to examine the technical background against which this discovery was made. Since the expansion of trade by sea in the sixteenth century, British dyers had at their disposal a fairly wide range of colouring matters derived from plant or animal sources. Reds could be obtained from the Kermes insect, the cochineal beetle and—in the form of madder—from the plant *Rubia tinctorum*. Yellow, sacred to Imperial China, came from the saffron crocus, from weld, fustic and quercitron, and blues were obtained from woad and indigo. Purple (which meant a range of shades from dark blue to deep red) came originally from the yellow juice of a Mediterranean shellfish, undergoing aerial oxidation to the colour which came to be synonymous with wealth and position. Most of these colours could be modified by the use of iron, tin or aluminium salts; at first these were used in such common forms as alum and copperas (Chapter One) but late eighteenth-century chemical technology made available the acetates of these metals.

Industrial archaeologists who search old directories will often encounter the trade description 'cudbear'. This dye was a product of Scotland and the name is a corruption of Cuthbert. In 1758 George and Cuthbert Gordon began to make the dye for colouring linen, silk and wool in imitation of the shades obtained from indigo. A lichen (several species would serve) was cleaned, dried and bruised in an ammoniacal liquor and finally treated with lime. The ammonia was obtained from human urine, collected in Edinburgh and Glasgow by men who were provided with the original Twaddle hydrometers to protect them against a water-down product. George and

Charles Macintosh (later to be the partner of Charles Tennant) are said to have paid £860 a year for such a supply of ammonia. Like the other vegetable dyes, cudbear could be mordanted to a desired shade, from pink to blue, and the manufacture of this dye became a popular by-trade of the copperas makers who were close to the trade in mordants (Chapter One); in appearance and behaviour it was very similar to litmus.

The success of Perkin's dye was due particularly to its fastness to light. Of the vegetable colours, the blues and purples were notorious for fading, even one day in bright sunlight being often sufficient to discharge the colour. Mauve was not in fact the first synthetic dyestuff: in 1771 the eccentric recluse Peter Woulfe (of Woulfe's bottle) had prepared picric acid from indigo and nitric acid in his cluttered and barricaded rooms in Barnard's Inn.[2] Picric acid, however, had not been developed commercially, in spite of the startling yellow which was later to cause alarm among the munition workers at Lydd in Kent. Only in 1851 did the Manchester firm of Roberts, Dale and Co. make picric acid in sufficient quantity for silk dyed by that means to be shown at the Great Exhibition. There was also a small trade in murexide for calico printing. The firm of Robert Rumney extracted uric acid from Peruvian guano, and at the time of Perkin's discovery was converting the uric acid from twelve tons of guano per week into twelve hundredweight of murexide at a price of thirty shillings a pound.

Perkin's factory was established at Greenford Green near Harrow in June 1857. Here, too, he had a private laboratory, a facility that he was never to be without for the rest of his life. He later confessed that at this stage neither he nor any of his associates had ever seen the inside of a chemical works. Benzene of uncertain quality was bought for five shillings a gallon from Miller & Co. of Glasgow, and nitrated in large iron vessels. Nitric acid of sufficient concentration was not to be had so it was made *in situ* by treating the benzene with sulphuric acid and sodium nitrate. The reduction to aniline was carried out by Béchamp's process, then only three years old, using iron filings and acetic acid: this was a very violent reaction and needed careful control (and some courage) on the part of the men. Dichromate for the oxidation cost about elevenpence a pound. After a year or two the firm of Simpson, Maule and Nicholson was able to supply a satisfactory nitrobenzene, a step which led them into the dyestuff trade.[3]

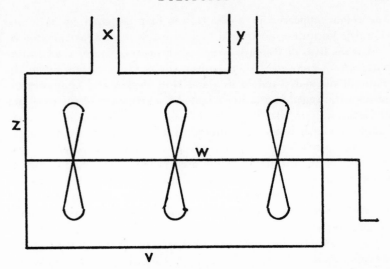

Figure 11 Perkin's Apparatus for Nitrating Benzene with Sulphuric Acid
and Sodium Nitrate, 1856

v. Cast iron vessel holding forty gallons
w. Stirrer turned by crank
x. Outlet for nitrous fumes
y. Inlet for benzene and acid
z. Removable end plate for emptying

In spite of all the technical difficulties, within six months of the
building of the factory mauve, or aniline purple, was in use for
dyeing silk at Thomas Keith's dyeworks in Bethnal Green. The
problems were not merely chemical. The dye possessed such an
affinity for fabrics that the portion of a length to be dipped first
acquired the deepest shade, and considerable research was necessary
before the idea of dyeing in a soap bath emerged. Perkin visited the
various kinds of establishments where his product was likely to be
used (as Muspratt had done a quarter of a century earlier), a calico
printing works in Dunbartonshire, a woollen business in Bradford,
and in each he spent weeks in collaboration with the local men until
the difficulties had been conquered. Perkin's genius is displayed in
this at least as much as in his chemical discoveries.

It was natural that Perkin's discovery should lead to a closer
examination of the industrial potential of aniline, and a second ani-

line colour appeared in 1859. It was first prepared by a French chemist, Verguin, and named fuchsine: the process, as operated by the Lyons firm of Renard, consisted in heating commercial aniline with stannic chloride. Other manufacturers made the dye under the name of magenta, and in England both Perkin and Hofmann examined the reaction and investigated the structure of the dye. One of Hofmann's pupils, E. C. Nicholson, was a partner in Simpson, Maule and Nicholson and he brought the manufacture of magenta to perfection under yet a third name of rosaniline, later changed to roseine. At first Simpson, Maule and Nicholson supplied roseine to the dyers in the form of an alcoholic solution, but later they were able to deliver a crystalline product, usually the oxalate. Rosaniline could be substituted by means of methyl, ethyl or phenyl groups to form the range known as Hofmann violets.

So far, all attention had been concentrated upon aniline or aniline derivatives, but in 1868 a new chapter in the history of synthetic dyestuffs was begun when Graebe and Liebermann in Germany demonstrated the possibility of making alizarin from anthraquinone. Alizarin is the colouring matter of madder, made from the ground roots of a plant first cultivated in ancient Egyptian times and by the eighteenth century well established in France, Spain and Italy. The German process was to convert anthraquinone into dibromanthraquinone and then to fuse with alkali. Perkin was at work on the same topic, and arrived at a method of preparation using fuming sulphuric acid followed by alkali fusion: in May 1869 he sent dyed patterns to the firm of Robert Hogg in Glasgow, and in June he took out a patent only to discover that Graebe and Liebermann had patented the same process a day earlier in Germany. However, so far as England was concerned there was no competition, and one ton was manufactured at the Greenford Green Works in 1869; by 1870 the make had risen to forty tons and by 1871 to 220 tons.[4] Very large quantities of caustic soda were used in this process, and Perkin noted that in Germany an organization such as the Badische Anilin und Soda Fabrik (BASF), bridging as it did the organic and inorganic chemical trades, could make its caustic on the site where the need for it arose. It can be remarked in passing that the growing dyestuff industry was a very large outlet for the older inorganic chemicals.

The consequences of the discovery of synthetic alizarin were disastrous for the madder producers. Prior to 1869 it had been estimated that 70,000 tons of madder root were cropped each year, of which about one-third came from France. In Britain alone, some 23,000 tons were consumed annually, the value being about £1 million pounds. Fourteen years after Perkin's plant was put into operation, no madder at all was being cultivated and it was said that in Smyrna there were large stocks still in the ground which the owners would gladly give to anyone who would carry them away, so that the ground might be used for other crops. We shall find this story to be repeated in the case of indigo.

The only sulphuric acid of the required strength that Perkin could buy was the old Nordhausen acid imported in large earthenware bottles: the danger and expense of transporting this acid caused a great deal of concern and led Perkin to seek a way of using home-produced acid. He discovered that anthracene previously converted to dichloranthracene would sulphonate easily in ordinary hot oil of vitriol, and that the product could be heated to yield hydrochloric acid, sulphur dioxide and anthraquinone disulphonic acid. For the alkali fusion, he first used a revolving cylinder heated in an air-bath, small iron balls being introduced to mix the products. At last he came to the method of heating in a strong iron boiler, under pressure—a kind of plant that was soon adapted to a number of dyestuff processes.

This device, known as an autoclave, had its origins in the 'New Digester for Softening Bones' brought out in 1681 by Denys Papin, a Frenchman who had been assistant to Christian Huygens and who worked with Robert Boyle in England. The lid was kept down by means of long screws, and the pressure was adjusted by moving the balance weight on a safety valve. Papin was able to prepare from old dry beef bones a palatable jelly 'as stomachical as if it had been jelly of hartshorn', thus demonstrating the increased power of solution of water boiling at elevated temperatures under increased pressures.[5]

Perkin, who had caused so much chagrin to Hofmann by retiring from pure chemistry to become an industrialist at the age of eighteen, now retired from industry at the age of thirty-six to devote himself once more to pure chemical research. In 1874 he sold the Greenford Green works to Brooke, Simpson and Spiller, the successors to Simpson, Maule and Nicholson with whom he had had close and early connections: with the factory went the licence to manufacture under patents owned by the Badische Anilin concern. These rights later passed to Burt, Bolton and Haywood and ultimately to the British Alizarin Company. This Company was formed with capital provided by British calico printers and Turkery-red dyers in answer to a threat of higher prices imposed by the German manufacturers.

Fortunately for the future of the British chemical industry, a few years before Perkin retired from business there had appeared on the scene a second giant of the dyestuff trade. In 1864 Ivan Levinstein set up his dyestuff factory at Blackley, Manchester, and for the next fifty years his prophetic voice exhorted government and industry alike to an awareness of the danger besetting the colour trades.[6] Levinstein's particular target was the loophole in the English patent laws which allowed German firms to take out patents in this country with no intention of working them, but with the aim of making Britain dependent upon German supplies: largely due to his perseverance, the Patent Act of 1907 incorporated a compulsory working clause. These activities brought upon him the hostility of the big German enterprises, from whom he had previously bought intermediates, and led to his making his own factory completely independent.

The Blackley factory at first made magenta, the various violets, Bismarck brown, chrysoidine and the famous 'Blackley blue': as the manufacture of intermediates grew, and the range of colours widened a new site was developed at Crumpsall in 1887. Bismarck brown and chrysoidine were early examples of azo-dyes, containing the grouping $-N=N-$ and in a different class from mauve and the violets. The reaction upon which the manufacture of these colours is based had been discovered in 1858 by Peter Griess, who became Hofmann's assistant at the Royal College of Chemistry and subsequently entered the employment of Allsopp's at their Burton-on-Trent brewery: it is of interest that he was drawn into this firm (where he stayed for twenty-six years) in order to refute a French allegation that pale ale owed its bitter flavour to strychnine. Griess, a sad-faced man of eccentric dress and mannerism, showed that nitrous acid could convert coal-tar bases into highly reactive intermediates which in turn could form coloured products with phenols.

Griess had neither technical perceptions nor commercial aspirations, and he received no financial reward for his discovery although it opened up completely new fields for others to harvest. The first

successful azo-dye was discovered in 1863 by Martius, working for
the Manchester firm of Roberts, Dale and Co., by the action of
nitrous acid on metaphenylene diamine: this was Manchester, or
Bismarck, brown. The second, chrysoidine, was produced in 1876 by
O. N. Witt at the Brentford factory of Williams, Thomas and Dower.
In 1884 H. Bottiger discovered Congo Red, an azo-dye with the most
valuable property of dyeing wool and cotton directly without the
application of a mordant: this discovery marked the beginning of a
highly productive period in the manufacture of these dyes. At one
stage in the preparation of such a dye, it was necessary that one of the
two components should be kept cool, a requirement which gave rise
to the name 'Ice colours', associated with the Huddersfield firm of
Read, Holliday and Sons. Thomas and Robert Holliday observed in
1880 that cotton might be impregnated with one of the components
and then passed through a bath containing the second, when an azo-
dye would be produced in the fibres of the cloth.

One of the consequences of Levinstein's victory in the matter of
German patents in this country was the establishment of a German
factory at Ellesmere Port on the River Mersey for the manufacture
of indigo. This dye (which is closely related to the woad of the Ancient Britons) was extracted by a complicated process from a plant
grown chiefly in India: in the last quarter of the nineteenth century
it was estimated that the value to India of this export was £4 million.
In 1868 Adolph Baeyer had begun a series of researches into the
constitution of this dye, but it was not until 1897 that Heumann
worked out a commercially feasible synthesis based on phthalimide,
and therefore ultimately on naphthalene. The Badische company
spent about a million pounds in research on the translation of this
process from the laboratory to the factory, and during the years of
their struggle the price of natural indigo fell to such a level that it
hardly seemed worth proceeding with the synthetic method. Between 1901 and 1907 new processes were introduced, based on
phenylglycine which in turn is produced from aniline. The German
factory in England worked successfully until 1914 when it was placed
under the direction of an official operator. By that date, however, the
whole question of German influence in British dyestuff consumption
had come under review.[7]

Sir Henry Roscoe had said in a lecture to the Royal Institution in
1881: 'To Englishmen it is a somewhat mortifying reflection, that

whilst the raw materials are produced in our country, the finished and valuable colours are nearly all manufactured in Germany.' Five years later Professor Raphael Meldola carried out a survey of the purchasing patterns of a number of dyeing and printing establishments and revealed an alarming situation. Edward Ripley and Sons of Bradford, perhaps the largest dyers of piece goods in the United Kingdom said that they used 86½ per cent foreign and 13½ per cent English coal tar colours; Walter Walker and Son of Dewsbury, dyers of wool for rugs, mats, carpets and blanket stripes, used at least 80 per cent German dyes ('to our own interest and advantage we have to do it'). Manson and Henry of Glasgow, yarn dyers, used only German dyes ('for both cheapness and quality'), and Heys and Sons of Barrhead said that of 10,000 pounds of colours only 700 pounds were of English manufacture. A firm of agents in London similarly affirmed that of colours passing through their hands, 95 per cent were German.[8]

The position was anomalous to say the least. Great Britain had been the birthplace of both the coal-tar products and the dyestuffs industries, many of the major developments in both had taken place here and several brilliant German chemists had felt it worth while to seek employment here; furthermore, the supporting heavy inorganic chemical industry owed its health and prosperity largely to British inventions. Yet the dyestuff industry was undoubtedly in thrall to Germany. Several reasons have been advanced for the failure of the industry to live up to its early promise: the excise duty and restrictions on the use of industrial alcohol, the import duties into foreign countries, the weaknesses of patent laws were all invoked to explain an otherwise inexplicable situation. There were some who blamed the lack of prestigious technical institutions, pointing to Zurich and Karlsruhe, and the absence of any encouragement for the young industrial chemist to prolong his studies: others held simply that the captains of industry did not devote enough of their resources to research.

The captains of the dyestuff industry, however, were themselves the innovators and experimenters: and Roberts, Dale had employed Caro and Martius as Williams, Thomas and Dower had employed O. N. Witt. Indeed a large proportion of Hofmann's brilliant band of students and assistants either founded dyeworks themselves or gained posts in established firms. Greville Williams worked at

Perkin's Greenford Green works before founding Williams, Thomas and Dower; Simpson, Maule and Nicholson, of the firm of that name were all Hofmann's students. What seems incontestable in this story is that the pioneers were not personally attracted to the world of expanding business. Perkin retired in 1874, a rich man, at the age of thirty-six, Nicholson, also wealthy, retired in 1868 aged forty-one, Greville Williams left industry in 1877 at forty-eight and Caro returned to Germany at the age of thirty-five in 1867.

The sheer weight of the German competition, once Hofmann's pupils had followed their master back to their own country, can scarcely be over-estimated. During the period 1886–1900, no less than 948 English patents were taken out by six of the largest German firms, namely Badische Anilin, Baeyer, Cassella, Mulheim and Leonhardt, Berlin Anilin, and Meister, Lucius and Bruning. In the same period only eighty-six were taken out by six English firms, Brooke, Simpson and Spiller, Clayton Aniline, Levinstein, Read Holiday, Clause and Ree, and W. G. Thompson. The emphasis on the patent literature was noticeable in the content of research. In 1887 Bernthsen (author of the textbook of organic chemistry which sustained several generations of students) left his university post at Heidelberg to join the research staff at Badische Anilin and found to his surprise that the only guidance that he was given was a list of patent specifications of competitor firms. In the early years of the present century, Carl Duisberg claimed that the great factory of Baeyer was testing as many as fifty new dyes in one day.

It is a curious fact that much of the earlier success in dyestuff synthesis had been achieved in an empirical way. In the great days of mauve there was, of course, little theory to guide the organic chemist: Kekule's ring formula for benzene came out only in 1865 and the condensed ring formula for naphthalene in 1866. Ten years later O. N. Witt propounded his theory of atomic groupings ('chromophores') which could impart colour to a compound, and 'auxochromes' which conferred dyeing properties. This enabled dye-stuff chemists to concentrate upon molecules containing such groups as $-OH$, $-NO_2$, $-NO$, $-N=N-$, with some certainty of partial success: partial, because only experiment could show if the combinations of shade, fastness and price would prove the birth of a winner. After 1888, when H. E. Armstrong drew attention to the prevalence of quinonoid structures in dyes, a good deal of work was done on the effects of conjugated double bonds upon the colour properties of dyestuffs.[9]

When the first world war broke out the chemistry of dye-making had reached a considerable level of sophistication while the health of the British dyestuff trade had steadily deteriorated until only about a dozen firms were still in business. Although they made a fairly wide range of dyes, magenta, malachite green, methyl violet, methylene blue, the indulenes and nigrosines and the sulphur blacks and greens, the intermediates were nearly all imported. The trade was truly described as 'threadbare'. Government intervention was the only possible solution to a national problem on such a scale. Consumers of imported dyes knew only the trade names and not the chemical constitutions of what they bought, and a Board of Trade survey revealed that no less than 10,000 names were in use; even allowing for trade synonyms, the sorting out of these was a major effort. Much of the plant, too, came from the Continent, particularly enamel pans, autoclaves and filter presses.

The Ellesmere Port indigo factory was sold to the bidder considered most likely to conduct the manufacture efficiently, and not surprisingly the choice fell on Levinstein's company. At the same time a small group of companies was formed round Read Holliday and Sons under the name of British Dyes Ltd. When the war was over British Dyes and Levinsteins amalgamated to form the British Dyestuffs Corporation. During the war James Morton, a textile manufacturer who specialized in fadeless fabrics, began to make the vat dyes which he had previously bought from Germany: his company was first called Solway Dyes and, after its move to Grangemouth, Scottish Dyes Ltd.[10] In 1925 the British Dyestuffs Corporation acquired Scottish Dyes, and a year later was itself merged into the great amalgamation with United Alkali, Brunner-Mond and Nobel Industries to form Imperial Chemical Industries Ltd.

The lot of the dyestuff chemist had been hard enough when the only fabrics with which he was confronted were wool, silk and cotton; for the affinities of these materials for colours varied widely. The introduction of synthetic fibres, however, set problems for which past experience with natural fibres was little guide. Viscose silk was essentially cellulose and so its peculiarities were quickly mastered, but the introduction of acetate silk was a very different matter. It was entirely devoid of affinity for dyestuffs of the ordinary kind, and

TABLE 7

FORMULAE FOR SOME HISTORIC DYESTUFFS,
SHOWING COMPLEXITY ACHIEVED

Mauveine 1856

Alizarin 1869

Fuchsine (Magenta) 1859

Chrysoidine 1876

Indigo 1890

Caledon Jade Green 1920

TABLE 8

PRODUCTS MANUFACTURED AT
PERKINS GREENFORD GREEN WORKS (1857–1873)

Mauve:	large quantities.
Dahlia:	Ethylmauveine, expensive and sparsely used.
Aniline Pink:	In washings from mauve.
Magenta:	Process based on mercuric nitrate proved dangerous and was discontinued.
Amidoazonaphthalene:	Scarlet pigment for calico printing.
Britannia Violet:	Large quantities.
Perkin's Green:	Extensively used in calico printing.
Alizarin:	Very large quantities.

Aniline salts, oxidising agents, and copper salts were also manufactured: coloured lakes for wallpapers, printing, and lithographic inks were made by processes which were Perkin's trade secrets.

so generally intractable that Morton called it the 'sphinx' of dyeing. In 1923, however, a range of 'Duranol' colours—water-insoluble dyes suitable for acetate rayon—was devised and in the following year there appeared the corresponding water-soluble dyes known as 'Solacet' colours.

Two other historic achievements of the British dyestuffs trade, now safely out of its doldrums again, must be recorded. In 1920 the research chemists of Scottish Dyes were engaged in the examination of derivatives of anthraquinone (the parent of alizarin) and succeeded in making a compound of extraordinary fastness and dyeing properties: this was Caledon Jade Green, the first really fast green dye. Other members of the same class followed, including Caledon Dark Blue and Caledon Ming Blue. The colours of green plants and red blood seem to have little in common with each other or with the dark blue protective ink which disturbed squids pump into the surrounding sea, but they possess, in fact, the same structural skeleton of organic bases grouped around a central metal atom, magnesium, iron or vanadium in the cases mentioned. In 1928 the chemists at Scottish Dyes (by then merged with ICI) noticed a persistent and unwanted blue discoloration in the phthalimide which they prepared

20 Picric acid nitration pots, Brooke's Chemicals, Halifax, 1915
21 T.N.T. washing plant, H.M. Factory, Queensferry, 1917

22 Preparation of aluminium using molten sodium, Alliance Aluminium Co., Wallsend, 1889

from phthalic anhydride and ammonia: this blue colour proved to be an iron compound roughly analagous to the natural pigments, and further research resulted in the discovery in 1934 of Monastral Fast Blue in which the iron was replaced by copper. As was to be expected, the first commercial phthalocyanine pigment was soon followed by others such as Monastral Fast Green. These dyes are characterized by a pronounced resistance to chemical attack. The ICI Alcian colours are ammonium derivatives of phthalocyanines specially prepared for dyeing cellulose: they were introduced in 1948.[11]

CHAPTER TEN

Miscellaneous Chemicals

Gunpowder

In a square in Freiburg in the Black Forest there is a monument to
Berthold Schwarz, around the four sides of which are depicted the
stages in his alleged discovery of gunpowder. In the village church in
Ilchester in Somerset there is a plaque to the memory of Roger
Bacon, the Doctor Mirabilis of the thirteenth century, on which he
too is credited with the discovery of gunpowder. There are many
examples in the history of science and technology of simultaneous
discovery, when the general store of knowledge has reached the
requisite level: there is no real ground, however, for believing that
either Black Berthold or Friar Bacon invented gunpowder. More
probably it evolved over many centuries from the various incendiary
mixtures used for warlike purposes by the Greeks and others. Certainly it was in military use in the fifteenth century and for mining
purposes by the middle of the eighteenth.

Now the making of gunpowder is not a chemical operation: it
involves a mere mixing and no chemical reaction takes place (at least
by intention). It does, however, call for the supply of three chemicals
—sulphur, saltpetre and charcoal, in specified chemical and physical
states, and as the last two were made in this country they have a
place in our story.

Charcoal

Wood is a rich chemical raw material from which can be obtained
potash, cellulose, tar, acetic acid, a range of solvents and charcoal.
Unfortunately there is no way of producing all of these from one
process, so a decision has to be made as to which product is of most
immediate importance: of course each of the products has a technological significance of its own, charcoal having been used as a fuel
(a use revived in the recent cult of the barbecue), as a chemical in

TABLE 9
PRIMITIVE CHEMICAL INDUSTRY BASED ON WOOD

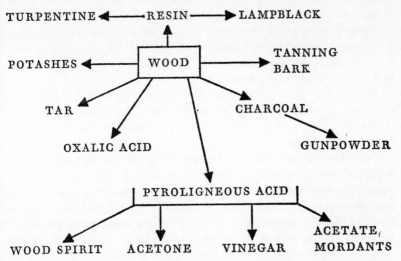

the production of carbon disulphide, as a reducing agent in such metallurgical processes as iron smelting, in addition to its use as an ingredient of gunpowder.

Like so many of the processes considered in this book, charcoal-burning had an indefinite origin; in many fairy tales the primitive charcoal-burner in the forest shares his benevolent image with his cognate technologist the wood-cutter. These early charcoal-burners made circular stacks of wood furnished with a central chimney and finished off in bee-hive shape with a covering of soil or turf. A slow combustion was maintained for perhaps five days and nights, control being exercised by means of wind-breaks and draughtholes. Only soft woods were used, the favourites being alder and willow, though Roger Bacon specified hazel twigs. As with the making of coke (Chapter Seven), the earliest methods for making charcoal contained no provision for the saving of byproducts.[1]

Byproduct recovery came in during the latter part of the eighteenth century; Bishop Watson claimed that he was responsible for introducing it to the gunpowder makers, and it may well be that there was some doubt about the qualities of charcoal from recovery plants—as there was about coke.

By the beginning of the nineteenth century, the process had settled into a standard form. Wood, previously freed from bark, was placed in large cast iron cylinders which were fixed horizontally in brickwork ovens and heated red hot by means of fires beneath them. The gases and vapours generated in the oven were led through traps to a condensing worm and thence to a receiver. When the charring was complete, the hot cylinder was lifted out with a block and tackle and a newly charged cylinder put in.

In addition to pyroligneous acid (an impure acetic acid), considerable quantities of tar were produced, described by Samuel Parkes in 1823 as 'a great burden on the hands of those who have been largely concerned in making the pyroligneous acid'. Parkes was in a position to know, for he had a contract to take all the pyroligneous acid produced by 'the King's cylinder works at Farnhurst in Sussex'. Some tar was used in treating timber for outside use, an outlet that was seriously challenged by Bethell's use of coal tar creosote in 1838. The pyroligneous acid was largely consumed by manufacturers of acetate mordants for the dyeing trade, to such an extent that—not for the first time in the chemical industry—the byproduct became more valuable than its parent chemical, and Parkes was able to state that 'those who may be desirous of having such charcoal may at any time procure it from the persons who manufacture either pyroligneous acid or gunpowder'.[2]

Saltpetre
Potassium nitrate was for a long time the bottleneck of the gunpowder trade. The related, and more plentiful, sodium salt suffered from the defect of deliquescence as a result of which the gunpowder became damp.

Until the sixteenth century saltpetre used in this country seems to have been imported from Spain, but under Queen Elizabeth we encounter the activities of 'saltpetre men' charged with the duty of going from place to place in search of ground from which saltpetre might be made: such ground would often be under pigsties and poultry runs. Conflict between the needs of industry and the amenity of citizens arose when the saltpetre men used their powers to invade private property. As late as 1848, Knapp's *Chemical Technology* described the appearance of a likely saltpetre site:

In densely populated towns, with narrow streets, where the excrements of beasts of burden, the refuse from slaughter-houses, and from trades of a like nature, where the water from the houses, the refuse of markets and other similar matters mix with the fluid in the drains, and are in a constant state of putrefaction, it may be seen how the coating of mortar at the base of the external walls is gradually eaten away, and becomes covered with a floccular, white, crystalline efflorescence.

Knapp warned that such an efflorescence should be tested to prove its authenticity, a useful property being its taste!

The powdermill at Waltham Abbey was producing gunpowder for the Government in 1560, and that at Chilworth in Surrey in 1570. There was then no hope of domestically produced saltpetre keeping pace with demand, in spite of the patent granted in 1561 for a 'statement of the true and perfect art of making saltpetre grown in cellars, barns etc, and in lime and stone quarries': not surprisingly, Neopolitan saltpetre was imported (along with the sulphur) at £3 15s per hundredweight. The East India Company began to bring in Indian saltpetre, and from about 1693 this material was used almost exclusively at Waltham Abbey.[3]

There was not, of course, much potassium nitrate in the earth which property owners yielded so reluctantly to the saltpetre men; lixiviation with water produced a solution of calcium nitrate which had then to be boiled with potash in order that saltpetre might be crystallized out. This led to an even fiercer competition for the supply of wood ashes, already in demand by glass and soap makers, and prepared from wood which was sought by shipbuilders, house carpenters, charcoal makers and—as a fuel—by almost every kind of manufacturer. The price of English wood ashes had begun to increase in 1636 when the saltpetre men offered two excuses for failing to reach their production targets: 'That wood ashes are so scarce, they can hardly be got at all. Those the petremen get cost $10\frac{1}{2}d$ or $11d$ a bushel which thereto once they got for $4d$. The unwillingness of most of the King's subjects to do anything for this service.' The pressure on wood ashes was not relieved until the Leblanc soda process (from 1791 onwards) provided glasshouses and soaperies with an alternative alkali.

The entry into Europe of Chile nitre, from 1830, brought a vast increase in the supply of nitrate, but linked—unhappily for the gunpowder makers—to sodium and not to potassium. Again a double decomposition was necessary, but after 1863 supplies of German carnallite (potassium chloride) began to take the place of wood potash for this purpose. The final product was known as 'conversion salt-petre'. By this time the supremacy of gunpowder was threatened. At Waltham Abbey, where nothing but black powder had been made for three centuries, the manufacture of guncotton was introduced in 1872. Ironically the new explosive was made in buildings which had earlier formed part of the saltpetre refinery. Cotton waste was treated with a mixture of three parts sulphuric acid and one part nitric acid at a temperature of 70°C, carefully washed and dried by centrifugation: it can therefore be seen that the process would not have been possible without the imports of Chile nitre from which the nitric acid was prepared. It is significant in this context, that as the British dyestuff trade declined (Chapter Nine) the manufacturers of the newer explosives, particularly Nobel's factory at Ardeer in Ayrshire from 1871, became the prime users of concentrated nitric acid (Pls. 20, 21).

Pigments

One of the earliest exercises in analytical chemistry is embodied in a direction by Pliny for detecting the adulteration of verdigris by the cheaper iron colours. The ancient world knew these blue and green carbonates of copper, the ochres of iron (later burnt to give umbers and siennas), cinnabar, azurite, malachite, orpiment and ultramarine (ground from lapis lazuli), and the Middle Ages added the mineral lakes and the smalt blue of cobalt. The large-scale manufacture of pigments (as opposed to the preparation by an artist of materials for his own use, as described by Benvenuto Cellini) began in Elizabethan times with the making of white lead.

In the lead-smelting areas such as Derbyshire, Yorkshire, Durham and Northumberland white lead was made by the stack process, in which strips of lead were built into a pile and exposed to acetic acid vapour and carbon dioxide: the former came from vinegar and the latter from horse dung and decaying vegetable matter. In 1787 Richard Fishwick of the Elswick Lead Works in Newcastle substituted for dung the spent bark from local tanyards: this factory still survives as part of Associated Lead Manufacturers, the chemical subsidiary of Goodlass Wall and Lead Industries Ltd. It was common for the manufacture of white lead to be conducted by women, who cast the thin lead sheets, arranged them in stacks and put in place the vinegar and spent bark.[4] The stack was built up in a lofty room as follows: first a layer of ashes, then a layer of spent bark about two to three feet thick, and on top of this a battery of earthenware pots, five inches deep, containing vinegar: on the pots was laid a layer of six sheets of thin lead and over these a covering of wooden boards. Each layer of pots, with the bark beneath and the lead above, was called a 'bed' and the stacks consisted of seven or eight beds arranged vertically to fill the room. About 1,600 pots and four tons of lead went to each bed so that the room had to be of stout construction.

After erecting a stack, the door would be closed and the room left untouched for twelve to sixteen weeks. When the process had run its course, the room was opened and the lead sheets extracted, the coating of white lead being broken off by repeated bending of the sheets. In the days before the control of industrial health became a matter for legislation, the hazards connected with the retrieval of the white lead must have been enormous. Many works had a doctor to inspect the women (who were supposed to take a bath once a week) but it was a common boast that they had 'dodged the doctor'. Even more hazardous to health was the old way of making red lead, in which a man stood before an open furnace and raked or stirred the molten lead for perhaps six hours until oxidation had taken place (Pl. 24).

The stack process, whatever fermenting material was used, was very slow, three or four months being required for the conversion of one stack. Quicker methods were sought, but such obviously simple processes as precipitating lead carbonate from solution gave pigments with far less covering power than the older product. In 1898 Cookson (one of the founders of Associated Lead) began to use a chamber process in which the gases could circulate about the lead more freely than was possible in a stack. The lead pigments, particularly red lead, supplied not only the paint trade but also the manufacture of the lustrous lead crystal glass.

White lead encountered competition from the zinc and barium pigments during the nineteenth century. Zinc oxide had been known to the alchemists as 'philosopher's wool', but its preparation as an artist's pigment dates from the 1840s when Winsor and Newton started to manufacture it. In districts where sulphide vapours in the atmosphere were sufficient to turn white lead black, the appeal of zinc oxide paint was obvious, but its long term weathering properties were poor. A more successful zinc white for general purposes was discovered in 1861 and patented in 1874. This was J. B. Orr's lithopone, a mixture of zinc sulphide and barium sulphate in which the dense white of the zinc sulphide is considerably enhanced by the fineness of the barium sulphate. The pigment was originally made in Glasgow but after the destruction of the works by fire in 1880 Orr moved to London: in 1896 he set up Orr's Zinc White Works in Widnes, acquired in 1930 by the Imperial Smelting Co. Ltd. Barium sulphate was prepared for use as a pigment by itself under the name of blanc fixe, chiefly as a by-product of two other barium-based industries to be discussed later.

Blue pigments in relics of the ancient world are usually hydrated copper carbonates (such as azurite) or ground lapis lazuli. This latter colour has been made under the name of ultramarine since 1828. In 1814 the director of the glass and alkali works at Saint Gobain had his attention drawn to a blue colour in the black ash furnaces, and he in turn brought it to the notice of the great French analyst, Vauquelin. Analysis showed the blue to have the same composition as lapis lazuli, and soon afterward the Société d'Encouragement offered a prize for a method of producing the pigment at a reasonable price. The problem was solved in 1828 in France and Germany simultaneously, and within a year the Royal Porcelain Factory at Meissen was producing ultramarine. The method was to heat together china clay (or some similar natural silicate), sodium sulphate and coal, sometimes with the addition of a little sulphur: the resemblance to what goes on in a black ash furnace is obvious. In England ultramine was manufactured by the sons of Isaac Reckitt of Hull: the firm had made various blues since 1840, but the production of ultramarine ('Reckitt's Blue') begun in 1883. At the beginning of the present century the manufacture of this pigment (sold for the purpose of combating the yellowing of fabrics in the wash) had assumed large proportions, other centres of the trade being in Bristol, London and Dorset.

Readers who search old maps for evidences of early chemical industry will often encounter names such as Blue House or Blue Works. Almost always these refer to the making of Prussian blue, widely produced in this country since the middle of the eighteenth century, but discovered in Berlin in 1704. In that year J. C. Dippel, the originator of a universal medicine known as 'Dippel's Animal Oil', collaborated with a manufacture of red lakes named Diesbach in making a new blue pigment: the ingredients, at that time, were potash, copperas and animal refuse. The process was officially published in England in 1724, but some blue was manufactured in Gateshead-on-Tyne at a much earlier date, by a German Jew who had probably picked up the secret in Berlin: the business was eventually taken over by Thomas Bramwell, one of the pioneers of nitrogen fixation in the nineteenth century.[5] It is interesting to recall that Prussian blue became a staple item in the armoury of food adulteration. Large quantities were shipped to Canton in the spring and autumn of each year for what was euphemistically termed 'facing' China tea: it was said that the Chinese would not drink tea treated in this way but 'the foreigners preferred it like that'.[6]

When the mechanisms for the production of prussiates of potash came to be understood, the use of slaughter-house refuse gave place to cyanides made by furnacing potash, coke and iron in the presence of air. With the purification of coal gas over iron oxide, gas works found themselves in possession of iron salts and some of the larger concerns, including the Gas Light and Coke Co., made Prussian blue as an outlet for their copperas, and their spent gas-lime which contained cyanides.

Apart from its use as a blue pigment large quantities of Prussian blue were mixed with chrome yellow to form the popular Brunswick green. Chrome yellow manufacture was pioneered in England by Andrew Kurtz, a native of Germany who received his chemical training in France and who, on settling in England in 1816, began the manufacture of soap in London. Kurtz's notebooks give precise and accurate details of his processes but are written in an ungrammatical mixture of English, French and German. In 1822 Kurtz was making potassium dichromate in Manchester: the price was initially five shillings a pound but gradually fell to eightpence. It seems clear that like many other good chemists who engaged in commerce, Kurtz had no head for business, preferring to conduct chemical experiments rather than to cost up accounts. From 1831 to 1838 he made chrome colours in Liverpool. His chrome yellow was used to paint the carriage of Princess Charlotte, with the result that for one or two seasons the colour was used by all the fashionable world.[7] When the editor of the *Lancet* launched a campaign against poisonous colours in food in

1855, it was found that yellow sugar confectionery was usually coloured with lead chromate.

Hydrogen peroxide and oxygen

We have referred to the use of barium salts in the production of lithopone and blanc fixe; we must now examine two other related industrial uses of barium minerals. Hydrogen peroxide was discovered in 1818 by Louis Jacques Thenard. With Gay-Lussac he had found that barium oxide could take up a further atom of oxygen when heated to a dull red heat in air, and that the resulting peroxide of barium reacted with dilute sulphuric acid to form what he called 'oxygenated water'. This laboratory reaction became the basis of the hydrogen peroxide industry, though not without the overcoming of several difficulties. In the reaction

$$BaO_2 + H_2SO_4 \rightarrow H_2O_2 + BaSO_4$$

the particles of barium peroxide became coated with insoluble barium sulphate unless careful means of agitation were employed. Other acids were excluded on the grounds either of expense or of possible reaction with the product. Hydrogen peroxide was in demand for bleaching wool and straw, and to a lesser extent for preparing ivory and bone for working into implements and ornaments.

The connection between hydrogen peroxide and wool caused Bernard Laporte to commence the manufacture of this chemical in Yorkshire in 1888; the application to straw bleaching led him to set up a factory at Luton in 1898 to serve the straw hat trade. Under the name of Laporte Industries, the firm acquired interests in acid-making and in titanium dioxide as well as taking over the old established alum business of Peter Spence and the fine chemicals establishment of Howards of Ilford.[8] Hydrogen peroxide is manufactured nowadays by the use of organic hydrogen carriers, such as 2-ethyl anthraquinol, which are alternately oxidized with oxygen and reduced with hydrogen.

Apart from the production of barium peroxide for hydrogen peroxide, barium oxide served another industrial purpose during the latter part of last century. When heated to dull red heat (about 540°C), as already mentioned, the oxide is converted to the peroxide; when the temperature is raised to a bright red heat (about 850°C) the peroxide breaks down once more into the monoxide and oxygen.

$$2BaO + O_2 \rightleftharpoons 2BaO_2$$

The brothers Brin patented a process for making oxygen by means of this reaction and in 1886 Brin's Oxygen Company was founded with plants in most of the large manufacturing towns. The way of working was later modified so that the temperature was kept constant at 650°C and the pressure of air alternately raised and lowered. The Brin concern assumed the title of the British Oxygen Co. Ltd, and in 1906 began the manufacture of oxygen by Linde's process of fractional evaporation of liquid air.

Phosphorus

For three centuries the production of phosphorus has been a peculiarly English trade, though the element was not an English discovery. In 1669 a physician in Hamburg, Hennig Brand, known contemptuously in his own day as Dr Teutonicus, was engaged in a lengthy distillation of human urine mixed with sand. In this he might have been guided by the alchemistic doctrine of signatures, according to which Nature indicates her treasures by means of shapes and colours which the dedicated observer must learn to interpret; a heart-shaped leaf, for example, or a flower appearing like an eye, would lead to the discovery of cures for heart disease or sore eyes. Similarly, a yellow liquid of natural occurrence might have been held to point to the secret of making gold. Brand, however, did not find gold, but a yellow-white waxy substance which glowed eerily in the dark. Readers who live near Derby should inspect the painting by Joseph Wright of that place, in which Brand's discovery is anachronistically, but dramatically, portrayed.

The news of his discovery spread rapidly through Germany and the Low Countries, and in 1677 we find the new element displayed at the court of Charles II in London. From that time, Robert Boyle interested himself in the 'cold fire' or 'noctiluca', and in 1680 he perfected a process which he disclosed to the Royal Society. Boyle had an 'operator' named Ambrose Godfrey Hanckwitz who soon scaled up Boyle's process, and from the laboratory in Southampton Street, Covent Garden, phosphorus was sold to 'the curious' at fifty-shillings an ounce: a good deal was sold on the Continent where it was known as 'the English phosphorus' or 'Boyle's phosphorus'. This laboratory, the birthplace of one of the earliest chemical export trades, was described as 'opening with glass doors into a garden which extended as far as the Strand. His laboratory was a fashionable resort in the afternoons on certain occasions, when he performed popular experiments for the amusement of his friends.'

After the death of Hanckwitz in 1741 no great improvement was made to the manufacture of phosphorus until 1769 when Gahn and Scheele in Sweden discovered phosphoric acid in bones. Distillation of this acid with charcoal yielded phosphorus more cheaply than the urine process. For a time the trade passed into French and German hands, and in 1844 no less than £2,567 was spent in importing phosphorus. It was used for the manufacture of matches, the phosphorus variety being introduced in 1833 under the name of 'Turin Candles'; (the earlier friction lights and lucifers had not contained phosphorus). In 1845 Arthur Albright commenced the manufacture of phosphorus from bone ash (imported from South America) and by the end of the decade only £3 was spent on bringing in foreign phosphorus.[9] Albright journeyed to the countries of the Danube in search of bones from the beef-canning industry: he found what he wanted, but no one was willing to transport so noisome a cargo by river so he set up a calcining furnace near the site, sending bone-ash back to England.

The phosphorus was extracted by treating the bone ash with sulphuric acid, and subsequently distilling at white heat a mixture of charcoal and phosphoric acid, the product being collected under water. This process yielded white phosphorus, extremely prone to catch fire, and damaging to the health of the workers: some idea can be gained of the dreadful risks involved, when we remember that sticks of white phosphorus were cast in glass tubes, filled by workmen who sucked up the molten phosphorus by mouth. The general aspect of a phosphorus factory was described in *Chemical News* for 1861:

The long, yellow flames of phosphoretted hydrogen and carbonic oxide shooting forth from the escape pipes; bits of burning phosphorus spitting forth in fiery balls from little crevices or leaks at the mouths of the retorts; the incessant bubbling of the vapour of phosphorus and escaping gases in the basins of hot water; the almost unbearable heat of the furnaces on either side and from the red-hot flues under foot; the intolerable stench of phosphoretted hydrogen and burning phosphorus, combined to produce an impression on our senses which we cannot fail to remember.

In 1849 the Viennese chemist Schrotter, having previously discovered a way to convert the dangerous white phosphorus into the less poisonous allotrope, amorphous or red phosphorus, lectured on the subject to the British Association at its meeting in Birmingham. Albright saw in this a way of making safer the business of match manufacture, and quickly introduced Schrotter's process into his Oldbury factory. Although less reactive to air, red phosphorus easily detonates on contact with potassium chlorate, and one kind of hazard in the match factories was replaced by another. The solution to this problem lay in the 'safety' match, in which the chlorate was confined to the tip of the match. The red phosphorus was made from white by means of a plant of Albright's design in which some speed of reaction was sacrificed to greater safety.

The most significant change in the mode of phosphorus manufacture took place in 1888, when the old coal and gas fired retorts were replaced by electric furnaces. This was more than a mere change in the source of heat. The old retorts had needed charging, heating, cooling and scraping out for each batch of phosphorus made, whereas the new furnace could operate continuously: moreover, the initial treatment with sulphuric acid was no longer necessary, the rock phosphate being reduced to phosphorus by heating in intimate mixture with carbon and silica. This fundamental improvement was due to that pioneer of British industrial electrochemistry, Sir Richard Threlfall, who had much to do with the founding of the fused silica trade: these methods for the production of red and white phosphorus remain substantially unchanged today.

CHAPTER ELEVEN

Fine Chemicals

England in the early days of the Industrial Revolution was a paradise for the vendors of proprietary medicines. Mrs Montagu wrote from Italy: 'The English are easier than any other nation infatuated by the prospect of universal medicines, nor is there any country in the World where the doctors raise such immense fortunes.' Certainly the amateur in medicine enjoyed a great vogue, particularly if he bore the additional stamp of respectability marked by the cloth. One of the oldest compositions was Daffy's Elixir, made and sold by the Rev. Thomas Daffy, rector of Harby in Leicestershire, from 1673 and still popular two centuries later. This tincture of senna flavoured with aniseed was typical of a large number of herbal remedies which commanded large sales.

That large fortunes were to be made out of this kind of chemical enterprise can be seen from the case of Joshua Ward. The founder of the English sulphuric acid trade (Chapter Three) returned from exile in France in 1733 and began under royal patronage to push the sale and use of medicines bearing his name. His stock in trade consisted of a sweating powder (ipecacuanha and opium), a sweating draught (wine and opium), sundry pastes and powders and an emetic drop and pill. The last two captured the popular imagination and led to the circulation of such lampooning couplets as

> Before you take his Drop or Pill,
> Take leave of friends and make your Will.

Both in fact contained as their active ingredient the fused sulphide and oxide of antimony known as 'glass of antimony', well known as a violent emetic and purge. Yet Henry Fielding and Horace Walpole both expressed their obligation to Ward's treatment of their complaints. These preparations enabled Ward to live in style and to leave £5,000 in his will.[1]

The appendix to the Stamp Duty Act of 1804 (which placed a tax on proprietary medicines) contained a list of 453 preparations liable to duty. Some of these can, of course, no longer be identified, and their names would not have survived but for the accident of their being on sale when the Act was passed. The remainder, however, indicate the size of the trade in drugs and medicinals, and include such favourite remedies as Scotch Pills, (aloes, jalap and aniseed), Dutch Drops (turpentine, guaiacum, sweet spirit of nitre and cloves), Godfrey's Cordial (carraway, treacle, aniseed and opium) Hooper's Pills (aloes, myrrh and iron sulphate), James's Powder (antimony and bone ash), Friar's Balsam (tincture of benzoin), Singleton's Eye Ointment (mercury nitrate in lard) and Opodeldoc (an aromatic ammoniated soap).[2] It can be seen that a good deal of extraction, purification and chemical preparation must have gone on in the background of such a trade. There was also a considerable by-trade in the production of 'succedanea' or cheap substitutes for the more expensive drugs.

The substances from which the proprietary remedies were made included alkaloids (particularly of opium and cinchona), essential oils, bitter principles, gums, sugars, organic acids, soaps, together with mineral ingredients such as borax, calomel, Epsom and Glauber salts, bismuth and antimony compounds, magnesia, ammonia and sulphate of iron. The work of extracting active principles from vegetable sources called for solvents with widely differing characteristics. Alcohol, in the form of spirit of wine, had of course been known for hundreds of years, but the rise of organic chemistry at the beginning of the nineteenth century brought into being a range of synthetic solvents. Benzene has been described in Chapter Eight; it now remains to survey briefly five of the commonest solvents.

Carbon disulphide had been known by the name of 'sulphur alcohol' since 1796, when Lampadius had made it by distilling a mixture of pyrites and moist charcoal. The first marketable product was made by C. Brunner in 1829; by heating sulphur in graphite retorts he was able to make twelve to fourteen ounces in two hours. The large-scale production of this solvent in Britain dates from 1843. A vessel of sulphur was set at the base of a cast iron retort containing charcoal; when the retort was heated, sulphur vapour rose up through the red hot charcoal and the resulting carbon disulphide vapour was cooled in long condensing coils. The effect of this solvent on oil extraction has been discussed in Chapter Six.

Chloroform was prepared in 1831 by Liebig in Germany and simultaneously by Soubeiran in France. At first it was made by acting upon chloral with alkali, but the first commercial process involved distilling a mixture of alcohol and bleaching powder. Muspratt's description of chloroform manufacture (1860) called for 130 pounds of bleach and seven pounds of lime mixed to a paste with a little water; the paste was put into an earthenware still, twenty-five pounds of alcohol added and the vessel heated by means of steam. Alcohol was later replaced by acetone and the scale of operations increased until charges of eighty pounds of acetone and 800 pounds of bleach were common.[3]

In 1840 Regnault prepared carbon tetrachloride by the further chlorination of chloroform and in the same year Dumas made it from chlorine and methane, but neither process was commercially successful. Kolbe in 1843 discovered the method by which most of the solvent has been made since, namely by passing chlorine and carbon disulphide vapour through a red-hot tube.

Acetone was made by Dumas in 1832 by the action of heat on lead acetate (sugar of lead), but calcium salts were more generally used in this type of preparation. The most important source of acetone was wood; first, the destructive distillation of wood yielded acetone, wood alcohol and pyroligneous acid (an impure acetic acid), the latter being converted to its calcium salt and heated to yield more acetone. Ether had been known since the early years of the seventeenth century; from its mode of preparation from alcohol and sulphuric acid it was long known as 'sulphuric ether' and the term ether was applied to all fragrant volatile esters. In 1835 Dumas and Peligot described the preparation and purification of ether. Apothecaries' Hall in London had a large ether-producing apparatus consisting of a lead still with a pewter still head connected to a tin cooling tube: the metals were chosen as unlikely to break or to leak and so cause a fire.

Plant extracts were made by a number of old established concerns, among them Allen and Hanbury (founded in 1715 and now part of the Glaxo Group), Howard and Sons (founded in 1806 as a separate organisation and now part of Laporte Industries) and Thomas Morson and Son (founded 1821 and still in business at Ponders End, Enfield). The latter firm made alkaloids in 1821, and were the first English producers of opium alkaloids. The sale of opium products reached

23 Westminster Gas Works, 1843
24 Stirring red lead, Locke Blackett & Co., Newcastle, 1844

25 Manufacture of sodium, Wallsend, 1889

enormous proportions during the nineteenth century; apart from the official preparations such as laudanum and paregoric, there was a multitude of compositions for quietening fractious children, sold loose from back-street grocers' shops under such names as 'Nurse's Friend', or 'Infant Preserver'.[4] Clearly, the first step to the regularizing of the indiscriminate sale of opium compositions was the replacement of crude opium by pure standardized alkaloids.

By the middle of the nineteenth century there had appeared a class of fine chemicals to rival the vegetable nostrums. This newer and simpler group was known as health salts or fruit salts, and was from the outset susceptible to more rigid chemical control. The trade in health salts grew out of a rebirth of interest in the watering places during the eighteenth century. A hundred years earlier, cowherds on Epsom Downs had noticed that their cows would not drink from a certain spring, and on trying the water for themselves they found it to be bitter and laxative. In 1695 Nehemiah Grew, a physician from Coventry and a pioneer of plant physiology, prepared from Epsom water a salt which he called *sal anglice* or Epsom Salt. Three years later he took out a patent for its manufacture, and this Epsom Salt is the first true patent medicine: a company was formed in 1700 and a factory opened at Shooters Hill to work Grew's patent.[5]

The widespread use of Epsom salt in self-medication gave rise to a number of cases of poisoning, not because of the physiological properties of the salt but because of its physical resemblance to oxalic acid. Readers of *Pickwick* will remember the proprietor of a chemist's shop who tried to obtain exemption from jury service on the ground that his assistant was particularly prone to confuse these two chemicals. Sometimes oxalic acid was clearly labelled on the packet but the recipient was unable to read and so took the deadly poison for the mild remedy; in other cases the pharmacist's assistant was unable to read the label on the bottle and was misguided by the colour and shape of the crystals. No less a person than the Rev. John Wesley found it necessary to publish a simple (but not foolproof) test for distinguishing between the two.

The immediate social consequence was that watering places increased in popularity. The effect was felt not only in such famous spas as Bath, Cheltenham, Tonbridge and Harrogate, but hundreds of small village springs began to be endowed with powers verging on the miraculous. An eighteenth-century treatise on the use of mineral waters internally and externally listed sixty-eight diseases which might be cured by the waters; these included epilepsy, melancholy, diabetes and dropsy. Not everyone who might benefit from the waters could afford to visit a spa and take a lengthy course of treatment, and so to spread the trade along a wider front attempts were made to sell bottled waters in the towns. Some temporary success was achieved, particularly during the cholera epidemics, but no British product had the wide sale of, for example, Vichy or Spa waters on the Continent. The next step was to evaporate the waters and sell the residues in the towns, so that seller could reconstitute his own mineral water. A simple extension of this principle was to analyse the water, identify the salts, and mix the appropriate ingredients from the dispensary shelves.

In 1857 a druggist named Alfred Bishop had the idea of incorporating unpalatable medicines into effervescing granules: a typical formula contained citric acid, sodium bicarbonate, and Epsom salt.[6] These preparations became very popular and were used for administering iron and quinine: at the present time about twenty-two preparations are on sale in which this formula is used. One of the first was Eno's Fruit Salt, made by James Crossley Eno in Newcastle in 1852.[7] Eno is described in directories as a chemist, but he never qualified by examination: he lived to the age of ninety-five and made many benefactions to the Infirmary in his native town. Another early and continuing example of this trade, is Andrews Liver Salt, made about eighty years ago by two shopkeepers in Newcastle named Scott and Turner, and now produced by Winthrop Laboratories.

In the fourth quarter of last century people began to ask for the effervescent granules without the active drug, and there sprang up a new craze in pleasant summer drinks in which the medicinal salt was replaced by sugar and flavouring. The lemon-flavoured variety swept the country under the names of Lemon Kali, Al Kali or Sherbert, the advertisement often suggesting an oriental origin. Since this material was often sold in farthing lots by grocers, cheaper substitutes for citric acid were sought. The more reputable makers used tartaric acid, but inferior preparations contained potassium hydrogen sulphate under the trade euphemisms of 'Tartaraline' or 'Citrolene'. Lemon Kali was first made by the firm of Godfrey and Cooke, and the *Chemist and Druggist* for 1899 describes the quantity consumed as 'enormous'.

A new concept came into the trade in medicinals with the introduction of synthetic drugs not known in nature. The tyranny of opium as a pain-killer was broken by the marketing in 1877 of acetanilide under the name of Antifebrin. This simple aniline derivative was bought by the ton for making up into headache powders, the most famous being Daisy Powders. The use of this compound was later restricted, and it was put on to Schedule I of the Poisons List, on account of the sensitivity which many people showed to its breakdown products. The discovery of aspirin in 1893, and its manufacture by the German firm of Baeyer at Leverkusen, brought on to the market a second and safer synthetic pain-killer. Since the beginning of the century the pattern of approach to the synthesis of new drugs has become standardized. First the chemical structure of numerous natural products is established, and correlations sought between structure and biological activity. Then molecules are synthesized to contain structural features likely to produce desired physiological effects. At the same time the barriers between heavy chemical and fine chemical manufacture began to become less obvious.

Phenol, for instance, was undoubtedly a heavy chemical and a product of the coal tar industry. But its use by Lister in 1865 to cleanse the atmosphere in his operating theatre has led to the synthesis of purpose-built phenolic molecules possessing enhanced antiseptic properties. Moreover, many of the newer disinfectants contain chlorine, itself a product of the heavy chemical trade. A similar crossing of earlier boundaries is evident in the productions of acriflavine (a bright yellow antiseptic) in 1916, not by a pharmaceutical firm but by British Dyes Ltd and the Pharmaceutical Division of ICI Ltd. clearly draws upon the resources of a firm originally founded to make alkali, dyestuffs and explosives.

A new market for fine chemicals was opened by the photographic experiments of Daguerre and Fox Talbot about 1839. Silver salts, gelatine, potassium iodide, collodion and pyrogallol were all in demand. Hypo for fixing had been made as a byproduct of one of the sulphur recovery processes in connection with Leblanc alkali; here, too, the problem was one of purifying compounds already available rather than of seeking new methods of preparation. One of the pioneers in the production of photographic chemicals was John Johnson. The business was originally that of assaying gold and silver, but in 1829 pharmaceutical chemicals were being made. In 1915 the firm moved to Hendon, by which time photographic chemicals were a speciality. During the 1914 war it became obvious that aerial reconnaissance was to be of great value, and that photographs from the air would reveal enemy troop concentrations. Unfortunately, as in the case of the dyestuffs, Metol, amidol, hydroquinone and other developers were all imported from Germany. The matter was made more urgent by the increasing use of X-ray photographs in surgery. Johnson's took over the manufacture of the much-needed fine organic chemicals and have continued to make them ever since.[8]

The demand for common chemicals in uncommon states of purity gave rise to a new branch of the fine chemicals industry. In 1866 the firm of J. J. Griffin put out a catalogue of laboratory equipment in which about 650 pure chemicals were offered for sale; they also advertised special collections of reagents for schools, agricultural colleges, colleges of medicine and for amateur adepts with the blow-pipe. The publication in 1862 of Mohr's *Lehrbuch der Titrirmethode* and in 1863 of Sutton's *Handbook of Volumetric Analysis* gave a powerful boost to the production of pure analytical reagents. The firm of Hopkin and Williams founded in 1850, began to supply chemicals for analysis and research: their service was all the more valuable because they published the standards to which their reagents were made and tests by which the purchaser could check them. This latter kind of information had previously been published to meet only the limited requirements of users of the British Pharmacopoia, and even then applied largely to materials prepared in the pharmacist's own laboratory. In 1934 Hopkin and Williams set up a joint committee with British Drug Houses (formed in 1908 out of a number of older firms) to compile a list of acceptable analytical standards of purity; out of this came the range of 'analytical reagents' or AnalaR chemicals. These standards were improved in the late 1930s to provide MAR or 'microanalytical reagents' to conform to the strict requirements of Pregl's microanalytical methods: hitherto only the German factories of Kahlbaum and Merck had satisfied this growing market.

A further development took place in the years after the first world war in the use of elaborate organic compounds for analytical purposes. The range of indicators for determining hydrogen ion concentration was broadened through the introduction of sulphone-phthaleins, and about 1927 the manufacture commenced in this country of specific organic reagents (e.g. oxine) for the detection and determination of metals. The publication by Hopkin and Williams of *Organic Reagents for Metals* did much to popularize this method of analysis.

In Chapter Eight we have referred to the production by C. B. Mansfield of 'essence of mirbane' and its use as a perfume in soap. The British fine chemical trade has for long made a speciality of the manufacture of perfumes, flavours and essences. The firm of Boake Roberts, established in 1869, applied what were then the very new resources of preparative organic chemistry to the synthesis of perfumes and fixatives which had previously been expensively extracted from plant or animal sources. Phenyl ethyl alcohol simulated the odour of roses as did ionone that of violets and these, together with musk xylene, heliotropine, citronellol and the common acetates and propionates, brought stable and cheap perfume into a whole range of consumer goods. W. J. Bush & Co. founded in 1851, was at first occupied solely with the making of flavouring essences: in 1886 Bush absorbed the much older firm of Potter and Moore, makers of toiletries. They were the first in Britain to make salicylic acid, and in 1910 they began the large-scale manufacture of vanillin, used immediately in the custard-powder trade, coumarin, the fragrant ingredient of Tonquin beans, following quickly as did piperonal, benzaldehyde, salicylaldehyde and anisaldehyde (the essence of aubepine). There was concern in certain quarters about the effects of the more powerful synthetic flavours, particularly in the 'fruit drop' type of confectionery. A. W. Blythe, the author of a famous and authoritative book on poisons, feared that the indiscriminate use of esters (then known as ethers) might anaesthetise the children who ate the sweets. Both Bush and Boake Roberts now belong to the Albright and Wilson Group.

The movement of chemicals out of the field of 'fine' production into the tonnage groups is a mirror of two kinds of change within the chemical industry, for it reveals how new uses are found for compounds which once were laboratory curiosities, and at the same time how new methods of manufacture make available on the large scale materials that were once obtainable only in small quantities. After all, sodium metal and aluminium were fine chemicals only a hundred years ago, and well in to the present century few pharmaceutical chemists would have predicted that hydrogen peroxide would be used to drive submarines. This is all particularly true of the elements once classed as 'rare'. Indium, for example, was nearly missed by Reich in 1863 because he was colour-blind and was detected by Richter as a thin blue line in the spectroscope, yet the current number of a chemical journal on the writer's desk advertises indium salts in tonnage quantities. The history of the chemical industry has many chapters yet to be written.

CHAPTER TWELVE

Where to Look for the Facts

It must be conceded that the chemical industry does not readily furnish material for the industrial archaeologist. Obsolete chemical plant has never inspired the same sentiments as veteran motor cars or locomotives, and, where detailed accounts have survived, the end of a once-prosperous works is too often summed up in the final phrase 'sold for scrap'. A works at Jarrow-on-Tyne was sold to a dry dock firm only on condition that the site was previously cleared, and this situation might well be typical. The waste heaps, too, have not been matters for local pride: the communities which engaged in costly litigation over loss of amenity were not likely to spend more money on the conservation of monuments to pollution. However, the recent trend towards the recovery and landscaping of derelict industrial land provides an opportunity which the student of the chemical industry should not neglect.

The various civil engineering and public works projects should be examined for the coloured strata that they reveal, from which fairly precise archaeological dating can be deduced as the following example shows. A site in County Durham (Friars Goose, east of Gateshead) was to be landscaped and turned into a sports area. The preliminary survey showed chalk to a depth of ten feet, and in an area where much dumping of limestone ballast used to take place this caused no surprise. The operation of the mechanical diggers brought to light. beneath the chalk, a very deep layer of black material, smelling of sulphide and thixotropic* to a degree that caused serious holdups in the work; red and yellow layers were also discovered. The site, in fact, was part of an old alkali works waste heap, and the black material was calcium sulphide waste from the soda extraction tanks. After 1882, when Chance's sulphur recovery process began to be worked,

---

* Thixotropic: quicksands are thixotropic, the solid mass turning liquid when it is disturbed.

the waste no longer contained sulphur and the upper layer of chalk accumulated (p. 45). The yellow stratum was arsenic sulphide, and indicated the introduction of a process for making arsenic-free sulphuric acid; the red layer was fine iron oxide from the recovery of copper in burnt pyrites. Similarly, the building of a new office block has revealed red and blue strata, evidence of an old colour works making Venetian red and Prussian blue. Black layers laid bare in the construction of a bypass road proved to contain manganese dioxide on the site of an old glass-house.

In most districts, the major problem is to discover what early chemical manufactures were conducted. Directories are of use, but caution should be exercised as names were not always deleted when a company ceased to operate. Moreover, the classification of a company depends very much on the policy of the compiler, and a firm may not be recognized under an unfamiliar heading: a case comes to mind of the same firm being listed variously (and correctly) under charcoal, lampblack, vinegar, turpentine distilling, bone-crushing and coke ovens. It is a good plan to ask at a reference library for directories at ten-year intervals, 1800, 1810 etc, and in this way to sort out quickly the longlived companies from the others. Maps showing names of factories need to be used with similar caution, for the data on which the map is drawn often pre-dates the publication by several years.

The Reports of the British Association for the Advancement of Science are often of great value, for many contain addresses on the state of local manufacture. Table 10 shows the towns where these meetings have taken place. Visits and excursions to local points of interest (including factories) were, and are still, arranged, and the local handbooks for these visits are invaluable. It is usual also to produce a scientific survey of the district which provides a contemporary account of industry: the more recent surveys do not contain so much chemical detail as the older examples. These are catalogued in libraries under British Association.

Since 1881 the Society of Chemical Industry, too, has held annual meetings (Table 11) in centres of chemical manufacture, and these, too, include works visits. Local handbooks for these are not usually so easy to trace, but the visits are often described in the *Journal of the Society of Chemical Industry* (since 1950 bound as *Chemistry and Industry*). The early volumes of this Society contain membership lists which often include business addresses.

Names of firms, and the towns to which they belong can be obtained from the Alkali Act register, contained in the Report of the Alkali Inspector for 1863 and 1881. The 1863 list is shown in Table 12: the 1881 list is very long indeed, because it contains firms outside the strictly chemical fields if they had some small section devoted to a registrable process: corporation gasworks figure very largely in this list, on account of their sulphate of ammonia plants.

Town councils were reformed in the 1830s and most of them have kept published minutes or proceedings since that date. These often mention petitions against nuisance, or applications from firms for permission to build on vacant land. Local record offices are the best sources of this kind of document, and they may also have leases and other deeds which sometimes have a map affixed. Such legal documents, of course, have the additional value of settling firm dates. Rating returns are likely to be available at the same source, affording information about growth. If a mortgage was raised for extension, there is likely to be a survey and inventory: bills of sale may also be available. Much of this depends upon the interests of the district archivist, but often to create a demand is the first step towards satisfying it.

The use of libraries calls for some comment. As with the physical relics, so with the documentation, and it must simply be accepted that the history of the chemical industry lacks the organized literature of coal mining, railways or shipbuilding. The research worker is thus thrown back upon his own initiative to a considerable degree, and he must therefore learn to choose paths of investigation which are likely to show reasonable returns for effort. Combing old newspaper files, for example, is a time-consuming process which, in the author's opinion, rarely yields commensurate results. It is far quicker to ask if the library keeps scrap-books of newspaper cuttings relating to areas or subjects: obituary cuttings are often kept and these should be searched for names appearing in the directories. In every case a specific enquiry is more likely to prove fruitful than one of a more general nature, as it gives the library staff a recognizable objective. To formulate specific enquiries, the directory drill outlined above is strongly recommended.

Almost every public library has some general material on the history of technology, but only a few have made a speciality of the subject. To name libraries in this connection would be invidious, but they can be traced from the subject index to Volume I of the ASLIB Directory (1968): look under the headings of 'Technology—historical', 'Companies—history', 'Chemistry—historical' and 'Science—historical'. The Library of the Science Museum in South Kensington provides on request lists of references on specific topics in the history of science or technology.

TABLE 10

ANNUAL MEETINGS OF THE BRITISH ASSOCIATION FOR THE ADVANCEMENT OF SCIENCE

Aberdeen 1859, 1885, 1934, 1963
Bath 1864, 1888
Belfast 1852, 1874, 1902, 1952
Birmingham 1839, 1849, 1865, 1886, 1913, 1950
Blackpool 1936
Bournemouth 1919
Bradford 1873, 1900
Brighton 1872, 1948
Bristol 1836, 1875, 1898, 1930, 1955
Cambridge 1833, 1845, 1904, 1938, 1965
Cardiff 1891, 1920, 1960
Cheltenham 1856
Cork 1843
Dover 1899
Dublin 1835, 1857, 1878, 1908, 1957
Dundee 1867, 1912, 1939, 1947
Durham 1970
Edinburgh 1834, 1850, 1871, 1892, 1921, 1951
Exeter 1869
Glasgow 1840, 1855, 1876, 1901, 1928, 1958
Hull 1853, 1922
Ipswich 1851, 1895
Leeds 1858, 1890, 1927, 1967

TABLE 10 *cont*

Leicester 1907, 1933
Liverpool 1837, 1854, 1870, 1896, 1923, 1953
London 1931*
Manchester 1842, 1861, 1887, 1915, 1962
Newcastle upon Tyne 1838, 1863, 1889, 1916, 1949
Norwich 1868, 1935, 1961
Nottingham 1866, 1893, 1937, 1966
Oxford 1832, 1847, 1860, 1881, 1894, 1926, 1954
Plymouth 1841, 1877
Portsmouth 1911
Sheffield 1879, 1910, 1956
Southampton 1846, 1882, 1925, 1964
Southport 1883, 1903
Swansea 1848, 1880
York 1831, 1844, 1906, 1932, 1959

*During the war years, shorter meetings were held in
London; for obvious reasons no guide-books were issued.*

TABLE 11

ANNUAL MEETINGS OF THE
SOCIETY OF CHEMICAL INDUSTRY
(*only meetings within the British Isles are listed*)

Aberdeen 1952
Birmingham 1891, 1907, 1917, 1930, 1955
Bradford 1903
Bristol 1918, 1960
Cambridge 1923
Cardiff 1934, 1965
Edinburgh 1894, 1916, 1927, 1948, 1968
Exeter 1939
Glasgow 1888, 1901, 1910, 1922, 1935, 1959

TABLE 11 *cont*

Harrogate 1937

Leeds 1895, 1925, 1957

Liverpool 1886, 1893, 1902, 1913, 1924, 1936, 1954, 1967

London 1881, 1883, 1885, 1889, 1892, 1896, 1900, 1905, 1909, 1919, 1926, 1931, 1940–47, 1951, 1956

Manchester 1882, 1887, 1897, 1906, 1929, 1949, 1964

Newcastle upon Tyne 1884, 1899, 1908, 1920, 1933, 1950, 1962

Nottingham 1890, 1898, 1914, 1932, 1953

Oxford 1961

Sheffield 1911

TABLE 12

ALKALI WORKS REGISTERED UNDER THE ALKALI ACT OF 1863

C. Allhusen & Sons, Newcastle.

J. & W. Allen

Frederick Allen, London.

Thomas Adkins & Co., Birmingham.

Boyd & Alexander, Dublin.

Bridgewater Smelting Works, St Helens.

William Barton & Co., Leigh.

Burnden Chemical Works, Bolton.

Borax & Alkali Works, Widnes.

Bonnington Chemical Works.

Henry Bury & Co., Church.

Richard Bealey & Son, Radcliffe.

H. Becker & Co., Chadderton.

T. Burnett & Son, Bill Quay, Newcastle.

T. Burnett & Son, Dunston.

William Blyth, Church.

Thomas Bramwell & Co., Newcastle.

TABLE 12 *cont*

Blaydon Chemical Co.
Carr & Hill, Widnes.
Chance Brothers & Co., Oldbury.
Carville Chemical Co.
Cook Brothers, Newcastle.
John Cook, Newcastle.
Conham Chemical Works.
Connah's Quay Chemical Co., near Flint.
Counterslip Sugar Co., Bristol.
Crosfield Bros. & Co., St Helens.
Thomas Dentith, Manchester.
Evans & McBryde, St Helens.
Everett Bros., Chelmsford.
John Farmer, Manchester.
Thomas Farmer & Co., London.
Thomas Fleetwood, Widnes.
Friars Goose Chemical Works, Newcastle.
John & Thomas Garrett, Wigan.
McGeachy & McFarlane, Glasgow.
Gaskell Deakin & Co., Widnes.
James Gibb, Bristol.
James C. Gamble & Son, St Helens.
Greenbank Alkali Works, St Helens.
Alexander Hope, junr., & Co., Glasgow.
Heworth Chemical Works, Newcastle.
William Howarth, Manchester.
William Hill & Sons, Liverpool.
John Hutchinson & Co. (No. 1.) Widnes.
John Hutchinson & Co. (No. 2.)
William Hunt, Aire & Calder Chemical Works, Castleford.
Harrison, Blair, & Co., Bolton.
Hazlehurst & Sons, Runcorn.
Irvine & Bryce, Glasgow.
Jarrow Hill Chemical Company.

TABLE 12 *cont*

ALKALI WORKS REGISTERED UNDER THE ALKALI ACT OF 1863

Jarrow Chemical Works.
William Jones & Co., Middlesbrough.
Andrew George Kurtz, St Helens.
William Jos. Kane & Son, Dublin.
Low Walker Chemical Co.
Thomas Lomax Jarrow.
William Liddiard, Stepney.
Lea Brook Alkali Works, Wednesbury.
Muspratt Bros. & Huntley, Flint.
Mersey Chemical Works, Widnes.
William McLiesh, Belfast.
J. McKenny & Co., Dublin.
Morgan Mooney, Dublin.
J. Montgomerie & Co., Glasgow.
John Metcalf & Sons, Manchester.
Marsh Alkali Company, Swansea.
J. Marsh & Co (Parr), St Helens.
James Muspratt, Liverpool.
J. Marsh & Co. (Ravenshead), St Helens.
Frederick Muspratt, Widnes, 2.
Netham Chemical Works, Bristol.
Netherfield Acid Works, Glasgow.
New Road Chemical Works, St Helens.
H. L. Pattinson & Co., Newcastle.
Port Tennant Copper Works, Swansea.
John Riley, Accrington.
Ebenezer Ridshaw Ridley, Newcastle.
Roberts, Dale & Co., Warrington.
Runcorn Soap and Alkali Works.
Richard Smith, Glasgow.
John Smith & Co., Bolton.
Seaham Chemical Works.

TABLE 12 *cont*

St Helen's Chemical Co.

Shaw & Hill, Widnes.

James Shorthouse & Sons, Birmingham.

Smith & Mawdsley, Flint.

Slack, Ashworth & Co., Liverpool.

Christopher James Schofield, Manchester.

Solomon Mease & Co., Jarrow.

Stevenson & Carlile, Glasgow.

Charles Tennant & Co., Glasgow.

Tennants & Co., Manchester.

Tyne Chemical Company, South Shields (lately St Cuthbert's).

Vivian & Sons, Swansea.

Weston Alkali Works, 2.

Edward Wilson, Prestolee.

Washington Chemical Works.

Walker Alkali Works.

George Whewell, Radcliffe.

John and James White, near Glasgow.

Select Bibliography

BIOGRAPHICAL

ALLEN, J. FENWICK, *Some Founders of the Chemical Industry*. Manchester, 1907. Allen was a copper smelter in Lancashire and his chemical detail is sound. The only book of its kind.

ALKALI

GOSSAGE, W. *A History of Soda Manufacture*. Liverpool, 1870. First-hand knowledge of a pioneer. Not much about other alkali centres.

KINGZETT, C.T. *History, Products and Processes of the Alkali Trade*. London, 1877. Essential reading.

LUNGE, G. *Sulphuric Acid and Alkali*. London, 1886, 1903. Lunge worked in England for 20 years before becoming Professor of Technical Chemistry at Zurich. Packed with comparative detail between British and continental practice.

RICHARDSON, T., and WATT, H. *Chemical Technology*. London, 1867, 3 vols. Richardson, alkali manufacturer and pupil of Liebig, draws much on his Tyneside experience. Much detail, and many drawings of plant.

DYESTUFFS

GARDNER, W.M. ed. *The British Coal Tar Industry*. London, 1915. In spite of the title, this is about dyes: authoritative papers on mauve, alizarin, indigo, patent laws, and the economics of dyestuff production.

Perkin Centenary. London, 1958. Chapters on the history of dyestuffs and of organic chemistry.

REGIONAL

BARKER, T.C., and HARRIS, J.R. *A Merseyside Town in the Industrial Revolution, St Helens 1750–1900*. Liverpool University Press, 1954.

Chapters on alkali, glass and copper.

CAMPBELL, W. A. *A Century of Chemistry on Tyneside*. London Society of Chemical Industry 1968. Alkali, glass, soap and colours.

CLOW, A., and CLOW, N. L. *The Chemical Revolution*. Batchworth Press, 1952. Comprehensive study of early chemical manufacture in Scotland. Very valuable bibliography.

HARDIE, D. W. F. *A History of the Chemical Industry in Widnes*. Widnes, ICI, General Chemical Division, 1950. Essential reading. Much use of company records in ICI archives.

HENDERSON, G. G. *Industries of Glasgow and the West Coast of Scotland*. Glasgow, Bristol Association, 1901. Chapter on chemical industries.

RICHARDSON, T. in *Industrial Resources of Tyne, Wear and Tees*, ed. W. Armstrong, Newcastle, British Association, 1863. Richardson was himself a chemical manufacturer.

GENERAL

HABER, L. F. *The Chemical Industry during the nineteenth Century*. Oxford University Press, 1958. Britain, Europe and America, from the economic viewpoint. Useful bibliography.

HARDIE, D. W. F., and PRATT, D., *History of the Modern British Chemical Industry*. Pergamon Press, 1966. Useful section on company developments.

MIALL, S., *A History of the British Chemical Industry*. Benn, 1931. Comprehensive company histories rather than chemistry.

MORGAN, SIR GILBERT, and PRATT, D. D., *The British Chemical Industry, its rise and development*, E. Arnold, 1938. Brief histories and contemporary details.

MUSPRATT, S., *Chemistry as Applied to Arts and Manufactures*. London 1860, 2 vols. Engravings of plant and factory interiors: much technical detail.

MUSSON, A. E., and ROBINSON, E. *Science and Technology in the Industrial Revolution*. Manchester University Press, 1969. Bleaching, dyeing and alkali. Careful documentation.

PARKES, S., *Chemical Essays*. London, 1823. 2 vols. Sixteen essays by a chemical manufacturer: 'chemical' widely interpreted.

COMPANY HISTORIES

ALBRIGHT AND WILSON: THRELFALL, R. E., *The Story of 100 years of Phosphorus Making*. Oldbury, Albright & Wilson, 1952.

BRUNNER AND MOND. *The First Fifty Years of Brunner and Mond.* 1873–1923. Winnington, 1923.

CASTNER KELLNER. *Fifty Years of Progress 1895–1945.* Birmingham, ICI, 1945.

CROSFIELD AND CO. MUSSON, A. E., *Enterprise in Soap and Chemicals.* Manchester University Press, 1965.

NOBEL INDUSTRIES. MILES, F. D., *A History of Research in the Nobel Division of ICI.* Birmingham, ICI, 1955.

TENNANT COMPANIES. TENNANT, E. W. D., *One Hundred and Forty Years of the Tennant Companies.* London, 1957.

THARSIS SULPHUR AND COPPER. CHECKLAND, S. G., *The Mines of Tharsis.* Allen & Unwin, 1967.

UNILEVER. WILSON, C. H., *The History of Unilever.* Cassell.

UNITED ALKALI CO. *Centenary of the Alkali Industry.* Widnes, 1923.

It is not expected that the general reader will have access to complete runs of bound chemical journals. For those with this sort of library facility, the most profitable journals for study are the *Journal of the Society of Chemical Industry* (1882), *Chemical News* (1860), *Chemical Trades Journal* (1888), and *Industrial Chemist* (1924). Anyone with leisure enough to browse through the volumes of the first named of these will find rich rewards. The following papers are classics:

CHANCE, A. M., The Recovery of Sulphur from Alkali Waste, *J.S.C.I.*, vi, (1887), 162.

GUTTMANN, O., The Early Manufacture of Sulphuric and Nitric Acid. *J.S.C.I.* xx (1901), 5.

LEVINSTEIN, I., The Development and Present State of the Alizarin Industry, *J.S.C.I.*, ii (1883), 213.

LEVINSTEIN, I., Observations and Suggestions on the Present Position of the British Chemical Industry, *J.S.C.I.*, v (1896), 351.

MUSPRATT, E. K., Survey of British Alkali Manufacture. *J.S.C.I.*, v (1886), 401.

MOND, L., On the Origin of the Ammonia Soda Process. *J.S.C.I.*, iv (1885), 527.

WELDON, W., On the Present Condition of the Soda Industry, *J.S.C.I.*, ii (1883) 3.

WELDON, W., On Some Recent Improvements in Industrial Chemical Processes. *J.S.C.I.*, ii (1883) 39.

The Penny Magazine, published by the Society for the Propagation of Useful Knowledge, ran a series of articles on factory visits which provide much contemporary detail. The following are worth examining:

Vol xi (1842), soap, p. 41; gas, p. 81; copper and lead, p. 249.

Vol xii (1843), alum, p. 421

Vol xiii (1844), chemicals, p. 201; glass, p. 249

Acts of Parliament are to be found in volumes entitled *Statutes of the Realm*. Each Act is dated from the year of accession of the Monarch in whose reign it is passed. Thus, since Queen Victoria came to the throne in 1837, an Act of 1839 is dated 2 Vict. The Alkali Acts are 26 and 27 Vict. 124 (1863); 31 and 32 Vict. 36 (1868); 37 and 38 Vict. 43 (1874); 44 and 45 Vict. 37 (1881).

Gazetteer

Two of the original alkali making areas have lost the monuments to this industry. The recent demolition of the St Rollox works with its memories of Charles Tennant has deprived Glasgow of its most important relic, and whilst certain small buildings remain in other parts of the city, multiple occupation has obliterated all useful signs of their former function. On Tyneside, too, all the old chemical sites have been occupied by shipbuilding or engineering enterprises. Several waste heaps remain, but plans for opening out the banks of the river to the public will certainly involve the removal of these unlovely memorials. Allhusen, Clapham, Losh, Tennant and Pattinson are remembered in street names (mostly in areas scheduled for redevelopment), and at Jarrow there is still the Alkali Hotel.

Elsewhere a number of historically important factories continue to operate, often making the product for which they were originally famed. However, the advances in chemical technology have caused them to change, often out of all recognition. They are nevertheless the sites on which industrial history was made, and merit the same respect as any other kind of historical location. Some of these are now listed: to visit them would constitute an act of piety rather than an exercise in industrial archaeology, but at least the imagination might be stirred.

ICI factories

Castner-Kellner, Weston Point, Runcorn: founded 1896 to make high-purity caustic soda by electrolysis of brine.
Dyestuffs Division, Grangemouth Works: originally Scottish Dyes founded 1920 by Sir James Morton: Monastral Blue and Caledon Green discovered here.
Nobel Works, Ardeer, Ayrshire: originally British Dynamite Co., 1871.

TABLE 13

SOME SURVIVING CHEMICAL FIRMS
OF EARLY FOUNDATION

1817	Johnson, Matthey and Co. Ltd.	London
1821	Thomas Morson and Son, Ltd.	Enfield, Middlesex
1826	Frederick Allen and Sons, Ltd.	London
1833	Eaglescliffe Chemical Co. Ltd.	Eaglescliffe, Co. Durham
1834	May and Baker, Ltd.	Dagenham, Essex
1837	Thomas Hedley (Procter and Gamble, Ltd)	Newcastle upon Tyne
1841	Lawes Chemical Co. Ltd.	Barking, Essex
1842	Turner and Newall, Ltd.	Washington, Co. Durham
1844	Albright and Wilson, Ltd.	Oldbury
1848	Burt, Boulton and Haywood, Ltd.	London
1850	Hopkin and Williams, Ltd.	Chadwell Heath, Essex
1851	J. C. Bottomley and Emerson, Ltd.	Brighouse, Yorkshire
1852	Reckitt and Colman, Ltd.	Hull
1863	Staveley Iron and Chemical Co. Ltd.	Chesterfield
1867	R. Graesser Ltd.	Sandycroft, Flintshire
1876	Clayton Aniline Co. Ltd.	Manchester
1877	Williams (Hounslow) Ltd.	Hounslow, Middlesex

The following combinations incorporate many old tar-distillation firms:

Bristol and Western Tar Distillers, Ltd.	Bristol
Lancashire Tar Distillers, Ltd.	Manchester
Midland-Yorkshire Tar Distillers, Ltd.	Oldbury
Scottish Tar Distillers, Ltd.	Falkirk
South Western Tar Distillers, Ltd.	Southampton

Research Laboratories: Widnes: descendant of the research laboratory of the United Alkali Co., founded 1892 and directed by Ferdinand Hurter.

Salt Works, Winsford, Cheshire: salt mine opened in 1844: triple-effect vacuum plant dates from 1905.

Billingham, Co. Durham: synthetic ammonia works set up after World War I to work Haber process: came into operation in 1923.

Imperial Metal Industries, Kynoch Works, Birmingham: in 1863 George Kynoch began here the manufacture of percussion caps.

Unilever factories

Lever Bros., Port Sunlight: William Hesketh Lever's first large-scale soapworks, 1888.

Joseph Crosfield, Warrington: founded 1814 by Joseph Crosfield of Lancaster: pioneers in the hardening of oils by hydrogenation and of oxygen-liberating washing powders.

Watson's Whitehall Works, Leeds: present premises acquired by Joseph Watson and two sons in 1860: Julius Lewkowitsch employed as chief chemist: long known locally as 'Soapy Joe's'.

Price's Bromborough Pool Works, Liverpool: originally candle makers: since 1854 manufacturers of fatty acids: among the first to recover and refine glycerol.

Laporte Industries

Laporte Acids, Hunt Works, Castleford, Yorkshire: a mid-nineteenth century bleaching powder works.

Peter Spence Works, Widnes: Peter Spence of Brechin began in 1845 to make alum from coal shale.

Albright and Wilson

Oldbury Works: established by Arthur Albright in 1851 to make white phosphorus: red phosphorus started in 1852: pioneers of electrothermal process 1896. Associated Chemical Companies, Mersey Chemical Works: old German (I.G. Farben) factory taken over by Brothertons in 1917: a reminder of the days when foreign firms worked their English patents in this way.

Associated Chemical Companies, Eaglescliffe Works, County Durham: established 1833 to make dichromates and other chrome products: later engaged in fertilizer trade.

Turner and Newall

Washington Chemical Works, County Durham: founded by Hugh Lee Pattinson in 1842 (later joined by R. S. Newall) to exploit his process for extracting magnesia from dolomitic limestone: site of first Glover Tower (1859), and of Isaac Lowthian Bell's aluminium plant (1859).

Monsanto Chemicals

Ruabon factory, Denbighshire: founded 1867 by R. Graesser to prepare phenol and cresols from crude tar acids.

Ciba and Geigy (joint owners)

Clayton Aniline Co., Manchester: established 1876 by Charles Dreyfus: coke oven benzole was refined to yield benzene, toluene and naphtha: benzene was converted to nitrobenzene and thence to aniline for the dyestuffs trade.

Among other firms who entered the interrelated fields of dyestuffs and tar distillation in the heyday of these trades may be mentioned: J. C. Bottomley and Emerson Ltd, Brighouse, Yorkshire: established 1851.
Burt, Boulton and Haywood Ltd, Silvertown Works, London E.16: founded in 1848, this firm acquired Brook, Simpson and Spiller who were successors to W. H. Perkin. Perkin always wished it to be known that he considered the Silvertown Works as the lineal descendant of the first synthetic dyestuff factory. Williams (Hounslow) Ltd., Hounslow, Middlesex; founded 1877 as Williams, Thomas and Dower: Greville Williams was Perkin's chemist at the Greenford Green Works.
The Science Museum at South Kensington has models of early chemical plant from which can be obtained a good impression of what an early chemical works looked like. Processes represented include Leblanc alkali (layout of a typical factory, with saltcake and blackash furnaces, lead chambers, Glover and Gay-Lussac towers, lixiviating tanks, and Weldon chlorine plant), Solvay soda, contact sulphuric acid, Portland cement, phosphorus, coal tar distillation, soap and dyes. The history of the fine chemicals trade is illustrated in the

Museum of the Pharmaceutical Society in Bloomsbury Square, London W.C.1., and in the Wellcome Historical Medical Museum, Euston Road, London, N.W.5.

When the open-air museum at Beamish Hall, County Durham is finished it will contain the pharmacy of John Walker of Stockton-on-Tees, who pioneered the manufacture of 'friction lights' in 1826.

Most industrial history is more easily investigated in company with like-minded enthusiasts. Lists of local history societies, including industrial archaeology groups, can be found in *Historical, Archaeological and Kindred Societies in the British Isles*, S. E. Harcup, University of London Institute of Historical Research, 1968. If after conducting enquiries along the lines suggested in this chapter, you conclude that far too little attention has been paid to the history of the chemical industry and its relation to our national well-being, the remedy is in your own hands.

References

CHAPTER ONE *Prelude to Industry: Salt and Copperas*

1. *Alkali Inspectors Report*, 1890, 27.
2. Pilbin, P. 'Industries of Newcastle and North-East England', M.Sc.Thesis, Durham, 1935.
3. *Victoria County History of Hampshire*, v, 1912, 469.
4. 'Imperial Chemical Industries Ltd: Salt Division', *Chemistry and Industry* 1954, 814.
5. Stewart, T. W., 'On the Tees Salt Industry', *J.Soc.Chem.Ind.*, 1888, 660.
6. Matthews, L. G., *History of Pharmacy in Britain*, E. & S. Livingstone, 1962, 239.
7. Hassall, A. H., *Food and its Adulteration*, 2nd ed, 1868–1968, London, 1876, 138.
8. Campbell, W. A., *A Century of Chemistry on Tyneside*, 1968, 8.
9. Clow, A. and Clow, N., *The Chemical Revolution*, Batchworth Press, 1952, 240.

CHAPTER TWO *Sulphuric Acid*

1. Campbell, W. A., 'Portrait of a quack: Joshua Ward 1685–1761', *Univ. Newcastle Medical Gazette*, June 1964.
2. Schofield, M., 'Variegation in vitriol, manufacture', *Chemistry in Britain*, 1967, 247.
3. Mactear, J., Address to Chemical Section, *Proc. Glasgow Phil. Soc.*, xiii, 1880, 409.
4. Clapham, R. C., *Trans. Newcastle Chem.Soc.*, i, 1868, 162.
5. Fleck, A., 'The British Sulphuric Acid Industry', *Chemistry and Industry*, 1952, 1184.
6. Aynsley, E. E. and Campbell, W. A., 'John Glover and the Clean Air Acts', *Chemistry and Industry*, 1959, 1540.

7. McDonald, D., *A History of Platinum*, London, Johnson Matthey, *from the earliest times to the 1880s*, 1960, 115.
8. Bell Papers in Gateshead Public library: Thomas Bell was a land surveyor and valuer who kept inventories of factories and estates.
9. Cook, Sir W., 'Peregrine Phillips, the inventor of the contact process for sulphuric acid', *Nature*, 117, 1926, 419.
10. Lunge, G., *Sulphuric Acid and Alkali*, 4th edn, I, iii, London, 1913, 1276.
11. Bedwell, W. L., *Production of Sulphuric Acid from Calcium Sulphate*, R.I.C. Lecture Reprints, 1952, No. 3.

CHAPTER THREE *Leblanc Alkali before 1863*

1. Clow, *The Chemical Revolution*, 65ff.
2. Gittings, L., 'The Manufacture of alkali in Britain, 1779–1789', *Annals of Science*, xxii, 1966, 175.
3. Clapham, R. C., 'An account of the commencement of the soda manufacture on the Tyne', *Trans. Newcastle Chem. Soc.*, i, 1868, 29.
4. 'Industrial celebrities: 3, James Muspratt', *Chemical Trade Journal*, 1889, ii, 240; Obituary of James Muspratt, *J.Soc.Chem. Ind.*, 1886, 314: Hardie, D. W. F., James Muspratt, *Endeavour*, 1955, 29.
5. Aynsley, E. E. and Campbell, W. A., 'Hugh Lee Pattinson, 1796–1858', *Chemistry and Industry*, 1958, 1498.
6. 'Industrial celebrities: 4, Christian Allhusen', *Chemical Trade Journal*, 1890, 222.
7. Campbell, W. A., *The Old Tyneside Chemical Trade*, University of Newcastle, 1964, 39.
8. Kingzett, C. T., *The Alkali Trade*, London, 1877, 102.

CHAPTER FOUR *Leblanc Alkali after 1863*

1. *Proceedings* of the Town Council of Newcastle upon Tyne, 9 Jan. 1839, 19.
2. Hodgson, G. B., *The Borough of South Shields*, Newcastle, 1903, 367.

3. *Alkali Inspector's Report*, 1864, 16ff.
4. Tennant, E. W. D., 'The early history of the St Rollox Chemical Works', *Chemistry and Industry*, 1947, 667.
5. 'Industrial celebrities: 1, William Gossage', *Chemical Trade Journal*, ii, 1889, 111.
6. Weldon, W., 'Recent improvements in industrial chemical processes', *J.Soc.Chem.Ind.*, Jubilee Number, 1931, 133.
7. 'Industrial celebrities: 2, Henry Deacon', *Chemical Trade Journal*, ii, 1889, 191; Hardie, D. W. F., 'The place of the Deacon Process in the history of chlorine manufacture', *Industrial Chemist*, 1951, 502.
8. Mond, L. 'On the recovery of sulphur from alkali waste', *Trans. Newcastle Chem.Soc.*, i, 1868, 75.
9. Chance, A. M., 'The recovery of sulphur from alkali waste by means of lime kiln gases', *J.Soc.Chem.Ind.*, Jubliee Number, 1931, 151.
10. Oliver, T., *Dangerous Trades*, London, 1902, 568.
11. Morrison, J., *Proc.Tyne Chemical Soc.*, 1873, 21 Nov.
12. Armstrong, W. G., *Industrial Resources of Tyne, Wear, and Tees*. Newcastle, British Association, 1863.

CHAPTER FIVE *More Recent Alkali Processes*

1. Lunge, G., *Sulphuric Acid and Alkali*, 2nd edn, III, London, 1896, 9.
2. Mond, L. 'The origin of the ammonia-soda process', *J.Soc.Chem. Ind.*, Jubilee Number, 1931, 143.
3. Miall, S., *History of the British Chemical Industry*, Benn, 1931, 196.
4. *Ibid.*, 182.
5. Pollitt, G. P., 'Twenty-five years of progress in the heavy chemical industries', *J.Soc.Chem.Ind.*, 1927, 291 T.
6. *Chemistry and Industry*, 1935, 411.
7. *Fifty Years of Progress, the Story of Castner Kellner Alkali Company*, London, 1945, 26.
8. *Handbook to the Industries of Newcastle and District*. Newcastle, British Association, 1889, 148.
9. Hardie, D. W. F. and Pratt, J. D., *History of the Modern British Chemical Industry*, Pergamon Press, 1966, 90.

CHAPTER SIX *Soap and Bleach*

1. McTear, J. 'The growth of the alkali and bleaching powder manufacture of the Glasgow district', *Chemical News*, xxxv, 1877, 23.
2. Checkland, S. G., *The Mines of Tharsis*, Allen & Unwin, 1967, 161.
3. Armstrong, H. E., 'Michel-Eugene Chevreul 1786–1889', *Nature*, 1925, cxvi, 750.
4. Gittings, L., 'The manufacture of alkali in Britain, 1779–1789', *Annals of Science*, xxii, 1966, 184.
5. Turner, E. S., *The Shocking History of Advertising*, Penguin Books, 1965, 83.
6. Musson. A. E., *Enterprise in Soap and Chemicals*, Manchester University Press, 1965, 11 ff.
7. Miall, *History of the British Chemical Industry*, 3, 240.
8. 'Industrial celebrities: William Gossage', *Chemical Trade Journal*, 1889, 111.
9. 'Unilever and the law: the Soap Trust libel', *Progress*, 1964, (2), 69.
10. Wilson, C. H., *History of Unilever*, Cassell, 1954, 2 vols.

CHAPTER SEVEN *Fertilizers*

1. 'In memoriam: Sir John Bennet Lawes', *J. Roy. Agric. Soc.*, 1900, 511.
2. Alford, W. A. L., and Parkes, J. W., 'Sir James Murray: a pioneer in the making of superphosphate', *Chemistry and Industry*, 1953, 852.
3. Muspratt, S., *Chemistry Applied to Arts and Manufactures*, London, 1860, ii, 563.
4. *Alkali Inspector's Report*, 1882, 74ff.
5. *Handbook to the Industries of Newcastle and District*, Newcastle, British Association, 1889, 155.
6. Gilchrist, P. G., and Thomas, S. G., *J. Iron and Steel Institute*, 1879 i, 120, and 1882 ii, 683: see also 'On the elimination of phosphorus in the Bessemer converter', *Trans. Newcastle Chemical Soc.* iv, 1877–80, 297.

CHAPTER EIGHT *Coal Tar Chemicals*

1. *Calendar of Domestic State Papers, Charles II*, 9 March 1681.
2. Clow, *The Chemical Revolution*, 389ff.
3. Everard, S., *The History of the Gas Light and Coke Company*, Benn, 1949, 17.
4. Lunge, G., *Distillation of Coal Tar*, London, 1882, 77.
5. *Royal Commission on Scientific Instruction*, 1872, i, 354 (E. Frankland's evidence).
6. Ward, E. R., 'Researches on coal tar', *Chemistry and Industry*, 1969, 1530; *Chemistry in Britain*, 1962, 373.
7. Mansfield, C. B., 'Researches on coal tar', *J.Chem.Soc.*, 1848, 244.
8. Mott, R. A., *History of Coke Making*, Cambridge, Heffer, for Coke Oven Managers' Association, 1936, 69.
9. Roscoe, Sir H., in Gardner, W. M., *The British Coal Tar Industry*, London, 1915, 106.

CHAPTER NINE *Dyestuffs*

1. Schofield, M., 'The founding of the coal tar dyestuffs industry', *Industrial Chemist*, 1956, 147.
2. Campbell, W. A., 'Peter Woulfe and his bottle', *Chemistry and Industry*, 1957, 1182.
3. Brightman, R., 'Perkin and the dye stuff industry of Britain', *Nature*, clxxvii, 1956, 815.
4. Perkin, W. H., 'On artificial alizarin', *J.Chem.Soc.*, 23, 1870, 133.
5. Wolf, A., *History of Science, Technology, and Philosophy in the XVI and XVII Centuries*, London, 1935, 548.
6. Levinstein, I., 'Observations and suggestions on the present position of the British chemical industry', *J.Soc.Chem.Ind.*, 1886, 351.
7. Haber, *The Chemical Industry during the Nineteenth Century*, 198: see also Gardner, *The British Coal Tar Industry*.
8. Meldola, R., 'The scientific development of the coal tar colours', *J.Soc.Arts*, 1886, 759.
9. Armstrong, H. E., 'Origin of colour and the constitution of colouring matters', *Proc.Chem.Soc.*, 1888, 27.

10. Cronshaw, C. J. T., 'In quest of colour', *Chemistry and Industry*, 1935, 515 and 547.
11. Paine, C., in *Perkin Centenary London*, London, 1958, 32.

CHAPTER TEN *Miscellaneous Chemicals*

1. *The Rise and Progress of the British Explosives Industry*, London, (International Congress on Applied Chemistry) 1909, 3.
2. Parkes, S., *Chemical Essays*, 2nd edn, London, 1823, i, 398.
3. *Journals of the House of Commons*, vols i–vi, 1547–1651.
4. 'A day at a lead works', *Penny Magazine*, xiii, 1844, 337.
5. *Industrial Resources of Tyne, Wear, and Tees*, Newcastle, British Association, 1863, 170.
6. Normandy, A., *Commercial Handbook of Chemical Analysis*, London, 1850, 550.
7. 'Industrial celebrities: 4, Andrew Kurtz', *Chemical Trade Journal*, 1889, ii, 287.
8. Hardie, D. W. F. and Pratt, J. D., *History of the Modern British Chemical Industry*, Pergamon, 1966, 405.
9. Threlfall, R. E., *The Story of 100 Years of Phosphorus Making*, Oldbury, Albright & Wilson, 1951, 8.

CHAPTER ELEVEN *Fine Chemicals*

1. Campbell, W. A., 'Portrait of a quack: Joshua Ward, 1685–1761', *Univ. Newcastle Medical Gazette*, June 1964.
2. Paris, J. A., *Pharmacologia*, 2 vols, 5th edn, London, 1822.
3. Muspratt, S., *Chemistry Applied to Arts and Manufactures*, London, 1860, ii, 470.
4. Blyth, W. A. and M. W., *Poisons, their Effects and Detection*, 4th edn, London, 1906, 292.
5. *Dictionary of National Biography*, viii, London, 1908, 609.
6. *Pharmaceutical Formulas*, (Chemist and Druggist), London, 5th edn, 1902, 503.
7. Campbell, W. A., 'James Crossley Eno and the rise of the health salt trade', *Univ. Newcastle Medical Gazette*, lx, June 1966.
8. Pyman, F. L., 'Twenty-five years of progress in the British fine chemical industry', *Chemistry and Industry*, 1935, 415.

Index